Posterior probabilities of alternative linear models

Some theoretical considerations and empirical experiments

Posterior probabilities of alternative linear models

Some theoretical considerations and empirical experiments

F.B. LEMPERS

Rotterdam University Press/1971

Printed in Belgium
ISBN: 90 237 2233 7

Foreword

An econometrician often faces the problem of choosing one model out of a set of possible models. If forecasting an endogenous variable is his goal, he may wonder which model predicts best. And if this task is advising a decision maker, he may look for the model which helps to make the best decisions. If the data available to him are scare or if he encounters multicollinearity (which in a sense amounts to the same thing), he may be forced to gamble. Then his position very much resembles that of a person whose goal it is to be the owner of a winning horse. If he can only afford to buy one horse, he is in the position of the classical hypothesis-tester who accepts one hypothesis and rejects all others. But if he is richer he can do better; he can keep a whole racing stable and enter for a race those horses which seem to be in the best condition at the moment. And from time to time he can replace certain horses that have a poor past performance and hardly better prospects.

An econometrician could follow a similar policy with his "stable of models". If he uses the Bayesian concept of the probability of a particular model being correct he may use mixtures of models for his prediction and decision problems. Assessing probabilities to models is not an easy job. The present book discusses a number of difficulties that arise and some ways of circumventing them. A final evaluation of these approaches will require extensive application and hence much time.

I have had the opportunity to discuss most of the material in the book at some length with the author. These discussions have given me great pleasure.

T. Kloek

Preface

The purpose of this book is to provide the applied research worker with theoretical and practical knowledge of the possibilities and difficulties of applying the concept of probability to discriminate between alternative linear models. For this purpose the reader needs some knowledge of Bayesian statistics, which is essential in order to read the book. The book is self-contained with respect to the concepts and techniques of the Bayesian statistics it uses. The reader is assumed to be familiar with calculus, linear algebra, probability theory and mathematical statistics.

I am most indebted to Professor Teun Kloek ,who suggested the subject and encouraged me in the research work that was needed to write this book. He taught me much of what I know of econometrics and Bayesian statistics, and the many hours he invested in discussing the early drafts of the book were immensely helpful. Without his support I doubt whether it would ever have been written. I am also indebted to Mr. Rinse Harkema of the Econometric Institute who made many valuable comments and was always ready to discuss parts of the manuscript. I am most grateful to Mr. A.S. Louter and Mr. F. van Doeland who gave valuable advice on writing the necessary computer programmes; and particularly to Mrs Cok Hempenius-Berger who typed the several versions of the manuscript so accurately. Finally, I wish to thank the Rotterdam University Press for their excellent printing. Needless to say, any errors that may remain are the author's sole responsability.

Fred B. Lempers

Contents

1. Introduction

1.1. MODELS

A large part of econometric work has been carried out to obtain numerically specified mathematical models that describe in a tractable and hence highly simplified manner what economists think about the relations in economic reality. These activities are undertaken for purposes of making predictions and helping to make decisions[1], or just to obtain a better understanding of the observed phenomena in question.

First, what are models? Orcutt[2] (1967), e.g., gives[3] the following description: 'They (models) are nothing but representations of something else. Typically they are representations in which details, that appear unessential for intended uses, are omitted'. In other words, a model is intended to represent the original object in certain significant respects, and is, therefore, a simplification of some original object. A model is easier to manipulate than the original object, since only the *relevant* properties of reality are represented. Orcutt (1967) argues:

> ... 'The advantage of a model over the real thing stems from the relative ease with which it may be manipulated or the low relative cost with which it may be created and experimented with ... .' ... 'For example, airplane models in a wind tunnel are a relatively low cost substitute for the flying of the real airplanes which are modelled. If, in fact, there exists a usable connection between the behavior of a model plane in a windtunnel and a real plane in actual flight, then a great deal can be saved in terms of lives, money, and time by carrying out experiments on and with such a model as substitute for equivalent experiments on and with an actual plane.'

1. MARSCHAK (1953) states in an introductory article of a book on econometric methods including parameter estimation of economic models: 'Knowledge is useful if it helps to make the best decisions'.
2. Throughout this book, references to items in the bibliography will be made by means of a year in parentheses following the author's name.
3. Most econometric textbooks do not pay attention to the principles of model building. MALINVAUD (1966), CHRIST (1966) and HUANG (1970), however, give some brief comments on these principles.

It is, of course, necessary to establish some correspondence between the behavior of models in wind tunnels and the behavior of real planes in flight, to obtain some idea about the approximateness of experimentation with such a model. Other well-known examples of models are, e.g., the use of the Link Trainer in the training of airplane pilots, the use of business games in the training of business executives, and the use of geographic maps such as city maps, ordnance maps, road maps, or weather charts. Concerning the geographic maps we can notice that nobody tries to present the information contained in all these maps on one map. The different maps are created for different purposes. We can say that map $\overline{M}$ is a model of the real thing M if $\overline{M}$ and M give similar answers to those questions considered relevant to the problem studied.[1]

Models, as we have already seen, may take different forms, e.g., verbal descriptions, graphs, (sets of) mathematical functions, physical embodiments (such as the aircraft model in the wind tunnel), and computer programs (e.g., Monte Carlo studies). Some of the models are physical reduced-scale representations of the original object, others are formal representations of the notions one has about a phenomenon (compare Malinvaud (1966), p. 49). In this book we shall stick to formal representations, more specifically, to models which describe real phenomena at the level of applied mathematics. Such models are called mathematical models, and are expressed by a set of mathematically formulated assumptions about the essential elements of the phenomenon and the laws which regulate it.

In the physical sciences a model can often be constructed in such a way that it is a faithful representation of reality. This is possible by the permanent nature of the underlying laws and, in particular, by the practical possibilities of experimentation, so that long sequences of observations become available. An important feature of the experiments carried out by physical scientists is that, in many cases, they are able to determine by means of a controlled experiment how two quantities vary simultaneously *ceteris paribus*.

The social sciences, however, differ in many respects from the physical sciences. We quote Malinvaud (1966) who argues:

> ... 'The situation is very different in the social sciences and particularly in economics. The phenomena which they study are bound up with the essentially moving organization of human communities. Experiment is often impossible, and great care must be taken in simply observing phenomena, as this can rarely be done under invariable conditions. The separate study of social and economic laws is therefore much more difficult than that of physical laws. The economist is forced to make simplified representations much more often than the physicist or the biologist. He must simplify not only when he is trying to make practical use of accumulated

1. Compare Toulmin (1960), Chapter 4, and Bertels and Nauta (1969), p. 168.

theoretical knowledge, but also when collecting or interpreting observations, and even when trying to formulate new theoretical relations.'

In economics several kinds of mathematical models can be created. We distinguish, for instance, between theoretical models and econometric models. We use the term *theoretical model* for a mathematical model which is derived by means of deduction from simple fundamental hypotheses, e.g., the hypothesis that every economic agent possesses and maximizes a preference function defined on an action space. Some examples of such deductions are, e.g., the theories of perfect competition, of monopoly, and of economically rational behavior by an individual. In many cases it is difficult to confront such a theoretical model with empirical observations, since the observations needed are not available.

An *econometric model* can be defined as a set of mathematical relationships (in some cases a single relationship) satisfying the following requirements: (i) the relationships describe aspects of economic reality, which are relevant to some specified problem; (ii) the variables in the relationships are observable; (iii) the mathematical shape is tractable (often linear); (iv) the number of unknown parameters is sufficiently small to make statistical estimation possible. (v) no *ceteris paribus* assumptions are made concerning variables which are not (approximately) constant over the set of units of observation.

It is conceivable that an econometric model is a simplification or approximation of a theoretical model. Therefore, economic theory can be an important source for the construction of an econometric model. On the one hand, economic theory can specify in some cases something about the algebraic signs and sizes of certain partial derivatives. It can also see to it that justice is done to certain homogeneity requirements, e.g., that all price and income elasticities of a demand function are adding up to zero as homogeneity of degree zero is required in the classical theory of consumer behavior. On the other hand, economic theory does not inform the econometrician how to retain the most relevant conceptual variables for the econometric model out of the generally large set of variables occurring in that economic theory. It also does not inform what empirical data should be used to represent those variables. Finally, it does not inform the econometrician about the shape of the functional relationship between the economic variables which appear in his model.

These considerations imply that the investigator is confronted with a situation in which several choices are available.[1] He can choose *one*

1. It is possible that no tractable model exists which satisfies all requirements imposed by theoretical considerations. Then, some requirements have to be dropped. This also is

reasonable form of such an econometric model on subjective grounds, including his personal experience, or he can try out each model out of the set of different reasonable models, confronting each one with relevant data, and then either choose one model out of the set of models, or decide to postpone the choice and to go on with several models simultaneously. Confrontation of the models with relevant data is often preferred to the subjective choice, since, given a decision rule and the class of models considered, it offers an objective decision.

1.2. LINEAR SINGLE-EQUATION MODELS

From now on we shall use the brief term 'model' meaning 'econometric model'. In this book we shall discuss a Bayesian procedure which enables the investigator to discriminate among (or choose from a set of) alternative linear single-equation models. Linear models are attractive because of their simplicity and mathematical tractability, and widely used in applied econometric work. Such a linear model is defined as an equation which is linear in unknown parameters. A typical representation is as follows

$$(2.1) \qquad y(t) = \sum_{i=1}^{r} x_i(t)\beta_i$$

where $y(t)$ is the variable to be explained (or dependent variable) for the t-th sample observation (or t-th observation period), β_i $(i=1, ..., r)$ are constant parameters, and $x_i(t)$ $(i=1, ..., r)$ the explanatory variables. The unexplained residuals of fitting a linear equation to observed data, for instance, by means of least squares, explicitly indicate the approximate character of explaining economic reality by means of the systematic part (that is the right-hand expression of (2.1)) of such a linear model. These residuals often show a rather capricious and unpredictable pattern, and, therefore, one often assumes that the model contains a disturbance term generated by a stochastic process. Then (2.1) can be reformulated[2] as

$$(2.2) \qquad y(t) = \sum_{i=1}^{r} x_i(t)\beta_i+\varepsilon(t)$$

a choice problem. For a more detailed discussion on this subject, we refer to SOMERMEYER (1967), for some additional remarks to KLOEK (1968).

2. Usually econometricians first derive a deterministic model and introduce the disturbances to complete the model. There are, however, also models where stochastic elements are incorporated from the beginning; see, e.g., CRAMER (1969), Chapter 2.

where $\varepsilon(t)$ is a random variable called disturbance. This disturbance can be considered as representing the neglected explanatory variables and is also a receptacle of all unpredictable elements in the behavior of individuals or institutions which have some influence on the variable to be explained. Finally, it may also contain errors of measurement of this variable to be explained. The randomness, in turn, means that a probability distribution on the disturbance term is assumed, which completes the specification of the model. Then the econometrician has the important tools of statistical inference at his disposal, which can be used inductively for getting information from a limited number of observations (called the sample[1]). It is therefore rather important what kind of assumptions are made about the form of this probability distribution, since in many cases such a hypothesis has a large influence on the final conclusions. For this reason econometricians are interested in testing their stochastic assumptions in the linear model.

We continue to investigate how such a linear model can be confronted with reality as observed by the sample. For this purpose, it is usual to write the general linear model in matrix notation, as follows

$$(2.3) \qquad y = X\beta + \varepsilon$$

where y is an n-dimensional column vector of random variables (called: sample observations), X and $n \times r$ matrix of explanatory variables with rank $m \leqslant r$, β an r-dimensional column vector of unknown parameters ($X\beta$ is often called the systematic part of the model), and ε is an n-dimensional column vector of nonobservable random variables: the disturbances. In textbooks on statistical methods of econometrics one often requires that X be a fixed (i.e., nonstochastic) matrix, but as will be shown in Section 2.4 this is not necessary for our purposes. The property of a fixed X matrix arises from the biological and agricultural sciences where experimental gardens, for instance, are used such that all circumstances which influence the experiment can be fixed by the investigator. This property is also mostly adopted by econometricians, although it is usually impossible to repeat the experiment with the same X matrix.

Usually, the following assumptions about the vector of disturbances are made:

(i) The mathematical expectations of the random variables $\varepsilon(1), \ldots, \varepsilon(n)$ are equal to zero; in vector notation

$$(2.4) \qquad E\varepsilon = 0$$

1. In cross-section analysis sampling techniques are frequently used. In time-series analysis, however, the term 'sample' often stands for a sequence of observations.

(ii) A positive definite symmetric (PDS) matrix Ω exists such that the variance matrix of the vector ε can be written as

$$E\varepsilon\varepsilon^t = \sigma^2 \Omega \tag{2.5}$$

where t denotes transposition. In many cases one assumes $\Omega = I$, where I is the identity matrix, which implies that the disturbances are uncorrelated and all have the same variance σ^2. (If $\Omega \neq I$, other simplifying assumptions on Ω are made to prevent the number of unknown parameters from becoming too large; see Chapter 4).

Sometimes, the following additional assumption about the disturbances is made;

(iii) The vector ε is a multinormally distributed random variable.

If one adds the requirement that X has full column rank (thus $m = r$) to avoid complete multicollinearity, then these assumptions provide sufficient additional information needed for estimating the vector β by means of least-squares or maximum-likelihood methods.

1.3. CORRELATION COEFFICIENTS

Apart from the problem of testing the assumptions about the disturbances, we encounter the problem of the choice of the X matrix. If the estimated parameters of the models considered have signs in accordance with theoretical considerations and are significantly different from zero, then (see, e.g., Goldberger (1968), Section 9.4) the highest squared multiple correlation coefficient (or determination coefficient) R^2 often plays an important role when choosing from a set of alternative linear specifications. This R^2 is defined by

$$R^2 = 1 - \frac{y^t M y}{y^t E y} \tag{3.1}$$

where $M = I - X(X^t X)^{-1} X^t$, $E = I - (1/n)\iota\iota^t$, and ι is an n-dimensional column vector with all elements equal to one. It is easy to verify that E is an operator which measures a variable as deviation from its mean. It can easily be shown that $1 - R^2$ equals the ratio of a *biased* estimator of σ^2 to the sample variance[1] of y. Therefore, R^2 may be corrected as follows

1. The term sample variance stems from the regression model, in which it is assumed that $y, x_1, \ldots, x_r$ are all random variables, which have a joint distribution. Then, if the marginal distribution of y has a variance, it is estimated by $\Sigma(y(i) - \Sigma y(j)/n)^2/n$; this

$$\text{(3.2)} \qquad \bar{R}^2 = 1 - \frac{n}{n-r}\frac{y^t M y}{y^t E y}$$

and this $\bar{R}^2$ is often used in practice. Concerning the relevance of this corrected multiple correlation coefficient, we quote Theil (1958, p. 211). He argues:

> ... 'The criterion of maximum multiple correlation is frequently based on rather primitive grounds: the investigator feels that the work he has done is of better quality if he can point out that he left only little of the variance[2] of his dependent variable 'unexplained.' Certainly, such a feature has a good deal of intuitive appeal, and efforts to find 'good' explanatory variables should be appreciated. It might even be argued that such a procedure has the virtue of maximizing the quality of the forecasts, for small residuals in the sample analysed are *prima facie* indicators of small prediction errors in the future. However, this argument is not very satisfactory, since it is far from sure whether the small residuals are due to the characteristics of the population or to the investigator's tenacity. Small errors in the sample are not indicators of plausibility, there being no fundamental law on economics stating that disturbances are small or 'as small as possible', nor are they indicators of simplicity.'

We feel a strong intuitive appeal to these arguments. Nevertheless, Theil obtains another justification for the application of the criterion of the highest multiple correlation for cases which are characterized by the same dependent variable. He derived under the afore mentioned assumptions (i), (ii) for $\Omega = I$, and the assumption that X is a fixed matrix, that this criterion leads 'on the average' to the correct choice; see Theil (1957), also Theil (1958), Section 6.2. Kloek (1970) showed under certain, not too restrictive, conditions on X matrices, that in the case of a sufficiently large sample the 'correct' model has the largest $\bar{R}^2$. It is worth noticing that if the assumption that the explanatory variables are nonstochastic must be replaced by the assumption that they are stochastic, then Theil's derivation is no longer applicable, and hence the criterion of maximum multiple correlation is not necessarily adequate on the average.

Up to now we excluded the cases in which the alternative specifications are characterized by different, but functionally related, dependent variables; in particular, those cases in which one specification has y and another $\log y$ (say) as left-hand variable. Theil (1957) shows that the comparison of a linear with a logarithmically linear relation by means of the correlation

quantity is called sample variance. There is much confusion about this term sample variance and the value of the same expression from the linear model. The remainder of this section also deals with other confusions between the regression model and the linear model.

2. See the preceding footnote.

coefficient is not straightforward, even if some additional assumptions are added. Another example is the case in which one specification is a linear transformation of another one. Especially in those cases where this transformation is not one-to-one, e.g., a level model versus the corresponding first-differences model, unsurmountable problems arise in showing that the criterion of maximum multiple correlation coefficient makes sense. Goldberger (1968), p. 130 also considers the shortcomings as mentioned before. He adds explicitly another warning, by saying: 'we should be wary of choosing functions which have a high R^2 but many 'nonsignificant' coefficients, or coefficients whose signs or magnitudes have no theoretical support'.

Another critical and interesting discussion of R^2 can be found in Koerts and Abrahamse (1969), Chapter 8. They point out the confusion among econometricians who define the statistic R^2 in the linear model and give it the interpretation as if it was defined in a so-called regression model. In a regression model the X-matrix consists of particular values of (normally distributed) random variables so that all variables are jointly distributed and generally there is no functional relationship between them. Then the R^2 can be interpreted as an estimate of an unknown parameter of the underlying multivariate probability distribution measuring the degree of linear dependence between two random variables. The differences between the regression model and the linear model seem obvious, although there is still enough resemblance between these two models to confuse practical workers. As we have already seen, the R^2 in the linear model is just a statistic describing the proportion of the total 'variance' in the dependent variable which is explained by this linear model. Following this train of thought Koerts and Abrahamse (1969), p. 130 make the following statement:

> ... 'In economic research, R^2 is often used in the quest for the correct specification of an economic law. Thus, from the class of theoretically acceptable specifications, that with the highest correlation coefficient is usually chosen. In such a case, the correlation coefficient is no longer used as a descriptive statistic but as a test variable for the correct specification of an economic law.'

They then go on to study whether or not it is possible to use R^2 as a test variable for the correct specification. After a considerable amount of mathematics they conclude:

> ... 'that values of R^2 belonging to different models can not be compared in a simple way since they have different backgrounds. Each coefficient should be interpreted in the light of its own probability distribution which, unfortunately, depends, apart from β and σ^2, on the specific values taken by the explanatory variables used in the model concerned.'

To solve this problem, Koerts and Abrahamse (1969), p. 152 suggest starting with *a priori values for* β *and* σ^2. This, in our opinion, stimulates a Bayesian approach of the problem of discriminating among alternative linear models.

1.4. ON THE BAYESIAN APPROACH IN THIS STUDY

In this book we sometimes make use of the expression the 'correct model'. Given our considerations in Section 1.1 it will be clear that no econometric model is correct (or true) in the sense that it is in full agreement with reality. The term correct model, however, could be used for the model that one is inclined to accept after mutual comparison with other models and confrontation with a large number of observations. In some cases a sample is generated by means of a computer in order to illustrate or investigate the properties of some method proposed. Then it is known which model was used to obtain the sample and it is obvious that this model is called the correct one.

One of the most challenging concepts in Bayesian statistics is the probability of a particular model being correct. Discussions of this concept can be found in Jeffreys (1961) and Carnap (1962). Some applications can be found, e.g., in Jeffreys (1961) and De Groot (1970). These authors are mainly concerned with testing hypotheses that certain parameters vanish. In other words, they use these probabilities for the purpose of significance tests. Geisser (1964) presents the probability that a particular observation belongs to one of k known multivariate normal populations, and Box and Hill (1967) developed a rule to select the experiment that attains maximum expected discrimination among k rival mechanistic models. Both approaches make use of Bayesian statistics, but apparently do not cover our econometric problem, since we neither have a number of different known multivariate normal populations at our disposal, nor are we generally able to carry out experiments in order to obtain a maximum of additional information.

More general attempts to apply the concept of probability to discriminate among linear models have been undertaken by Thornber (1966), Zellner and Geisel (1968), Geisel (1970), and Zellner (1971). A common point in the approach of these authors is that generally they accept to dispose of little or vague prior information and treat their problem as if they had no information at all. Thornber (1966) accepts the 'invariant' locally uniform 'diffuse' prior density, see, e.g., Jeffreys (1961), while the other authors also treat the problem of little information with a decreasing quantity of prior information and consider a limiting posterior distribution.

This book treats the same subject starting from natural-conjugate distributions, as introduced[1] by Raiffa and Schlaifer (1961). It is also reconsidered for the case of uniform prior distributions, since a wide-scale use of these prior distributions can be observed and hence it is useful to illustrate some of the difficulties inherent in this approach.

In principle, the posterior probability of a linear hypothesis could take over the role of R^2 in discriminating among alternative linear models. Many problems appearing in practice, such as testing different assumptions about the disturbances, lend themselves to rather natural solutions. Other problems, for example the choice between different explanatory variables, can not be solved straightforwardly. A large part of the book is concerned with these problems. The most important lesson one can learn from the present study is postponing the choice problem in case of doubt and going on with several models at the same time.

1.5. OUTLINE OF THE STUDY

The order of discussion is as follows. In Chapter 2 we briefly introduce some concepts of Bayesian statistics, necessary for the reading of this book. In Chapter 3 the so-called posterior probabilities are derived for natural-conjugate prior distributions, as well as for uniform prior distributions. In this chapter we also consider large-sample properties of these posterior probabilities and discuss a simple particular case. In Chapter 4 we apply the posterior probabilities to sets of linear models, differing only in the assumptions about the disturbances. In Chapter 5 we discuss how to compare models with different explanatory variables. Comparisons are made with the existing approaches of other authors. Finally, in Chapter 6, we present some illustrative experiments in which models, as discussed in Chapter 5, are compared. We also discuss how to compare models with different, but functionally related, variables to be explained.

We finish this introductory chapter with some technical remarks. The book is divided (as already mentioned) into six chapters, and each chapter into some sections. Sections and tables are numbered in the decimal notation. Section 4.2, for example, is the second section of Chapter 4. Formulas are numbered as follows: (3.5.1) refers to formula 1 of Section 5 of Chapter 3. Within Chapter 3, it will be referred to simply as (5.1).

1. Raiffa and Schlaifer (1961), in turn, refer to the work of Savage (1954) for a foundation of their approach. Later Pratt, Raiffa, and Schlaifer (1964) have given an alternative axiomatization of decision under uncertainty which is in essence simpler than Savage's.

2. Some concepts of Bayesian statistical analysis

2.1. INTRODUCTION; SOME BASIC CONCEPTS

It is about two centuries ago that Bayes' Theorem was published, not long after simple estimation procedures and significance tests had become available, which were the origins of many techniques of statistical inference which are now widely used. Since that time, there have been differences of opinion and many discussions about the interpretation and the validity of the methods which are nowadays called Bayesian. From approximately 1925 to 1955, the anti-Bayesian influence of R.A. Fisher was so dominant, that only a small number of Bayesian methods and a large number of non-Bayesian methods were developed during that period. Bayesian statistics was considered as a heresy by the 'orthodox' statisticians, and even the existence of this branch of statistics was mostly not mentioned in the text-books. 'The imbalance is now being corrected, and we can expect to see a greater emphasis on Bayesian activities in the next ten years than in the last ten', concludes Plackett (1966), who gives an interesting historical survey of the development of statistical inference.

Pioneering work of, e.g., Savage (1954) and Raiffa and Schlaifer (1961), changed the situation, and widely used statistical textbooks now pay brief attention to the Bayesian approach. See, e.g., Mood and Graybill (1963), Hogg and Craig (1965), and (for statistical methods of econometrics) Malinvaud (1969). For textbooks which give a much more detailed treatment we refer to Lindley (1965) and De Groot (1970).

For the reader who is not very familiar with the Bayesian train of thought, we start with a brief and informal review of some basic concepts of Bayesian statistics:

(i) *Bayes' Theorem.* Bayesian statistics is called Bayesian, because the application of Bayes' Theorem is essential. This theorem reads:

$$(1.1) \qquad P(A|B) = \frac{P(B|A)\,P(A)}{P(B)}$$

provided $P(B) > 0$. It is an elementary theorem in probability calculus. It is the interpretation Bayesians give to $P(A)$ and $P(A|B)$ which has been the subject of many discussions, that have not come to an end yet.

(ii) *Sample Space.* The term sample space may be used either to mean the set of all possible outcomes of a given experiment or (more widely) the set of all possible outcomes of all possible experiments with respect to a certain model. In many simple cases one means by all possible experiments: all possible sample sizes.

(iii) *State Space.* The term state space denotes the set of all possible 'states of nature' or 'states of the world' with respect to a certain model. In many simple cases the word 'state space' has the same meaning as the word 'parameter space'.

(iv) *Probability Distribution on a State Space.* Bayesian statisticians define probability distributions not only on sample spaces but also on state spaces; to be more precise: on Cartesian product spaces of sample spaces and state spaces. In doing so they give the word probability a wider meaning[1] than classical statisticians do, which is more in accordance with the meaning of the word probability in everyday speech. This meaning includes subjective probabilities as may be expressed by betting odds, but also much more objective probabilities, when long historical evidence is available.

(v) *Relevance.* The uncertainty that often bothers us in practice (in decision problems, but also in pure research) is uncertainty with respect to the 'true state of nature' rather than uncertainty with respect to the sampling properties of certain statistics. We are, e.g., interested to know how our market will react to our policy; not how, given a certain model and given a set of 'true' parameters, certain estimators of these parameters will behave. This makes Bayesian statistics interesting and relevant for applied work.

2.2. SOME MORE CONCEPTS

In this section we continue our discussion of some of the main concepts of Bayesian statistics, but on a somewhat more technical level than in the previous section. We closely follow Raiffa and Schlaifer (1961) and do not refer to their book on each occasion. On the other hand, we refer to other authors if necessary. We confine ourselves to prior and posterior distributions, since the main goal of our study is to find answers to the question what econometricians can do with a *given* body of data.

1. See, e.g., De Leede and Koerts (1970).

Essentially, Bayesian statistics is concerned with joint probability measures defined on a possibility space $\Theta \times Z$, which is the Cartesian product of a state space Θ and a sample space Z. The state space $\Theta = \{\theta\}$ is the set of all possible 'states of the world' which cannot be predicted with certainty. By means of so-called subjective betting odds, see, e.g., de Finetti (1937), Pratt, Raiffa, and Schlaifer (1964), and Raiffa (1968), it is possible to introduce the concept of probability. This enables the statistician to assign a distribution on Θ, which is called a *prior distribution.* Although the true state is unknown, it is possible to obtain additional information about the state. This can be done by some experiment. All possible outcomes of all potential experiments generate the sample space $Z = \{z\}$. Given some z one can revise his prior thoughts about Θ, if one can derive a distribution on Θ, given z. Therefore, one needs a joint distribution on $\Theta \times Z$, which implies the existence of the following distributions:

(i) a marginal distribution on Z;
(ii) a marginal distribution on Θ;
(iii) a conditional distribution on Z given θ;
(iv) a conditional distribution on Θ given z.

We note that 'classical' statistics (or sampling theory) studies distributions on Z, given some $\theta \in \Theta$. In a Bayesian context these distributions are conditional distributions; see (iii). Bayesian statisticians, however, are interested in conditional distributions on Θ, given some $z \in Z$; see (iv). For simplicity we confine ourselves here to the case that a probability measure on $\Theta \times Z$ can be described by a joint density function. Then, a marginal density on Θ is called *prior density*, and denoted by $D'(\theta)$; a conditional density on Z, given some experiment and some $\theta \in \Theta$ is called *likelihood* and denoted by $l(z|\theta)$; a conditional density on Θ, given some $z \in Z$, is called *posterior density* and denoted by $D''(\theta|z)$. These densities are related to each other by Bayes' Theorem (1.1) which implies under the conditions stated above

$$(2.1) \qquad D''(\theta|z) \equiv D'(\theta)\, l(z|\theta)\, N(z)$$

where $N(z)$ is the normalizing constant defined by the condition

$$(2.2) \qquad \int_{\Theta} D''(\theta|z)\,\mathrm{d}\theta = N(z) \int_{\Theta} D'(\theta)\, l(z|\theta)\,\mathrm{d}\theta = 1$$

We shall confine ourselves to prior densities D' and values of z such that the integrals in (2.2) exist and are positive. For practical purposes this is not a serious restriction.

Formula (2.1) is the basic formula for large parts of Bayesian statistical analysis. In words, it can be formulated as follows: the posterior density is equal to the product of the prior density and the likelihood, apart from a normalizing constant (which can be found by integrating the just mentioned product over the state space). Formula (2.1) can be simplified by deleting the normalizing constant. If D is either a prior or posterior density on Θ and K is a function on Θ such that

$$(2.3) \qquad D(\theta) = K(\theta)/\int_{\Theta} K(t)\,\mathrm{d}t$$

one writes

$$(2.4) \qquad D(\theta) \propto K(\theta)$$

and says that K is a (not 'the') *kernel* of D. If $l(z|\theta)$ is a likelihood on Z, and if ρ and κ are functions on Z such that

$$(2.5) \qquad l(z|\theta) = \kappa(z|\theta)\rho(z)$$

where ρ is independent of θ, one says that $\kappa(z|\theta)$ is a (not 'the') *kernel* of the likelihood $l(z|\theta)$. Accordingly, (2.1) may be rewritten as

$$(2.6) \qquad D''(\theta|z) \propto K'(\theta)\kappa(z|\theta)$$

The important consequence of (2.6) is that both in the prior density and in the likelihood we can drop all factors that are independent of θ, and compute the integration constant (which guarantees that $D''(\theta|z)$ integrates to unity) afterwards. It turns out that this saves much time and energy. Therefore, in practice one uses (2.6) much more often than (2.1).

The choice of a convenient prior distribution is not a simple problem. We distinguish between three cases: (i) The statistician represents his prior thoughts without any restriction. This approach usually requires the application of numerical integration methods; (ii) The statistician adopts some restrictions in order to make the prior distributions more tractable. Then he may make use of the so-called natural-conjugate families of prior distributions; (iii) The statistician states that he has no information *a priori*; this is often called vague prior knowledge, or ignorance. For a more detailed treatment of these three cases we refer to the next section.

Finally, we introduce the Bayesian definition of sufficiency. We shall assume that an abridged description y of the sample z exists and can be represented as a point in a Euclidean space Y such that the dimensionality of y does not depend on the dimension of z. We shall denote by $\tilde{y}$ the mapping which sends Z to Y. The event that the random variable $\tilde{y}$ assumes the value y is abbreviated as $\tilde{y} = y$. This mapping is not one-to-one. This means that different elements of Z can generate y. Thus y is just a coarser description of a particular experimental outcome and is called 'sufficient' if it contains all information about the sample that is required by somebody who applies Bayes' Theorem to find the posterior distribution on the state space. More formally, in Bayesian statistics a sample z can be described by a sufficient statistic y, if there exists a mapping $\tilde{y}$ which sends Z into Y and a mapping k which sends Y into R^+ (the set of nonnegative real numbers), such that

(2.7) $$k(y|\theta) \propto \kappa(z|\theta)$$

for all values of z satisfying[1] $y = \tilde{y}(z)$. As before, the proportionality sign indicates that factors which are independent of θ may be omitted.

Example. Consider an independent normal process with known variance 1 and unknown mean μ. The likelihood of a sample z of n observations on this process is

$$l(z(1), \ldots, z(n)|\mu) = (2\pi)^{-\frac{1}{2}n}\, e^{-\frac{1}{2}\sum_{i=1}^{n}(z(i)-\mu)^2}$$

$$= (2\pi)^{-\frac{1}{2}n}\, e^{-\frac{1}{2}\sum_{i=1}^{n}(z(i)-m)^2}\, e^{-\frac{1}{2}n(m-\mu)^2}$$

where $m = (1/n)\sum_{i=1}^{n} z(i)$, and the statistic $y = (n, m)$ is sufficient and of fixed dimensionality 2, independent of n. For we may drop all factors not containing μ, which results in

$$k(y|\mu) = e^{-\frac{1}{2}n(m-\mu)^2}$$

and this expression is a coarser description than

$$\kappa(z|\mu) = e^{-\frac{1}{2}\sum_{i=1}^{n}(z(i)-\mu)^2}$$

If $D'(\mu)$ is the prior density of μ, then the posterior density is

$$D''(\mu|m, n) \propto D'(\mu)\, k(m, n|\mu) = D'(\mu)\, e^{-\frac{1}{2}n(m-\mu)^2}$$

which is the same posterior density as results from combining $D'(\mu)$ and $\kappa(z|\mu)$. If the prior density is equal to

$$D'(\mu) \propto e^{-\frac{1}{2}n'(m'-\mu)^2}$$

then

$$D''(\mu|y) \propto e^{-\frac{1}{2}n''(m''-\mu)^2}$$

where[2] $n'' = n' + n$ and $m'' = (1/n'')(n'm' + nm)$.

1. Raiffa and Schlaifer (1961), pp. 32–35, give a more general definition, but the present definition covers all our and most other applications of the sufficiency concept. They also present a proof that Bayesian and classical definitions of sufficiency are equivalent in certain important cases.
2. An identical operation can be performed if one uses the sufficient statistics to combine two likelihoods. For more details, see Raiffa and Schlaifer (1961), pp. 44–47.

2.3. PRIOR DISTRIBUTIONS

In most applied problems the number of possible 'states' will be extremely large. An obvious way of specifying a prior distribution is to start by selecting some *family* of distributions defined by a mathematical formula containing a certain number of adjustable parameters and then to select a member of this family which meets the investigators' quantitative specifications of his prior ideas about the proper numerical values of these parameters. Raiffa and Schlaifer (1961), Section 3.1, discuss a number of desiderata which such a family of distributions should satisfy; of these we mention the following. It should be analytically tractable; this means, it should be reasonably easy to determine the posterior distribution resulting from a member of the family and a given sample. The family should be rich enough, so that there will exist a member capable of expressing the prior information. And the family should be closed in the sense that if the prior is a member of the family, the posterior will also be a member of the family.

Therefore, we now introduce the concept *natural conjugate*. It is possible to consider the kernel function k as a function $k(.|\theta)$ with parameter θ on the reduced sample space Y, and as a function $k(y|.)$ with parameter y on the state space Θ. If the integral of this function over Θ exists, a density $f(.|y)$ on the state space Θ with parameter y can be derived by defining

$$f(\theta|y) = N(y)k(y|\theta) \tag{3.1}$$

and where $N(y)$ is defined by

$$[N(y)]^{-1} = \int_\Theta k(y|\theta)\,\mathrm{d}\theta \tag{3.2}$$

The density (3.1) on Θ will be called a *natural conjugate with parameter* y of the kernel k; see Raiffa and Schlaifer (1961), p. 47.

Now, when the prior distribution is natural conjugate with a kernel of the likelihood we are able to derive the posterior distribution in a very simple way. To be more precise, the kernel of the prior density combines with the sample kernel in exactly the same way that two sample kernels combine; compare the example in Section 2.2. This implies that natural-conjugate prior distributions have the attractive properties of tractability and closure. For a more detailed discussion we refer to Raiffa and Schlaifer (1961), Chapter 3.

For the choice of a prior distribution out of the family of natural conjugate prior distributions we distinguish between some alternatives. If there exists a substantial amount of 'objective' (historical) evidence on which one has to base a prior distribution on the state θ, one can fit this distribution to these historical relative frequencies. If, on the other hand, no solid 'objective' (historical) basis for the assessment exists, the prior distribution has to be fitted to subjective betting odds with which the responsible person wishes his final decision to be consistent. In this case the psychological difficulty of assigning the total density is so great, since it will generally be impossible to specify the distribution in more detail than by a few measures such as the mode, or a few fractiles. But even then, if these specifications fit to a member of the family (or more members), one still has to study the eventual extreme implications of this density.

Econometricians, however, are often faced with a situation which is a mixture of the just mentioned extremes. Sometimes they have at their disposal short historical evidence or data of similar regions, so that some 'objective' evidence is available. For an illustration of such a case, we refer to Harkema (1971), Chapter 4.

Thus far, this section dealt exclusively with so-called natural-conjugate prior distributions. For completeness, we mention again the two other types of prior distributions.

(i) If an investigator has some definite prior distribution which is not a member of the family of natural conjugates, he still may apply (2.1). But analytical integration is then either more difficult or impossible. In the latter case numerical methods of integration are applied. For a more detailed discussion and application we refer to, e.g., Thornber (1967) and Chetty (1968). (ii) Many econometricians are reluctant to specify prior distributions, either since they believe that scientists ought to be 'objective,'[1] or since they dislike the difficult task of specifying numerically the rather vague information they have. There has been a great deal of discussion[2] concerning the representation of this state of mind by an appropriate prior distribution, and it seems rather unlikely that there will ever be general agreement. Raiffa and Schlaifer (1961) point out that the notion that a broad distribution of some θ is an expression of vague opinions is simply untenable, since the distribution of $1/\theta$ can be very definite. In other words, it is not always meaningful to distinguish between 'definite' or 'vague' prior distributions. A possible solution to this problem is that a prior distribution is called to be 'vague' if it can substantially be modified by a small number of sample observations. For a more detailed discussion we refer to Raiffa and Schlaifer (1961), Section 3.3.4. On the other hand, for large samples the

1. Formally, this concept is not well-defined. RAIFFA and SCHLAIFER (1961), pp. 68–69 state that scientists have to report in such a way that the user of their report has ready access to those aspects of the data which will enter into his own decision problem. An obvious solution to this problem is simply to report the data or, at least, the sufficient statistics. Many would-be 'objective' scientists do not satisfy this requirement.
2. For references, see PLACKETT (1966).

choice of the prior distribution is not very important. Plackett (1966), for instance, argues: 'In general, however, when the number of observations is sufficiently large (...) the likelihood will have a sharp peak at the maximum-likelihood estimate of θ, so that in forming the posterior distribution only a small interval of the prior distribution is relevant. Any prior distribution will be fairly constant in this range and therefore it can be argued that the sole criterion for selecting the prior is mathematical convenience.' For a small sample, however, the existence of some prior information can be very valuable. In such a case, it would be wasteful to neglect this prior knowledge.

In practice uniform prior distributions are widely used in the case of 'vague' prior information. Such a case can also be handled with a limiting posterior distribution obtained from a natural conjugate prior distribution. A more detailed treatment of these two possibilities, will be postponed until we have introduced the Bayesian approach to the linear model, see Sections 2.6 and 2.7.

2.4. NATURAL-CONJUGATE PRIOR DISTRIBUTION FOR THE LINEAR MODEL

We now apply the Bayesian approach to the linear model

$$(4.1) \qquad y = X\beta + \varepsilon$$

where y is an n-dimensional vector of random variables generated by (4.1), X an $n \times r$ matrix of explanatory variables, β an r-dimensional vector of parameters (real numbers, apart from their dimension) whose values remain fixed during an entire experiment, and where the elements of the n-dimensional vector ε are independent random variables with identical Normal densities

$$(4.2) \qquad f_N(\varepsilon_i|0, h) = (2\pi)^{-\frac{1}{2}} \, e^{-\frac{1}{2}h\varepsilon_i^2} \, h^{\frac{1}{2}} \qquad (i = 1, \ldots, n)$$

The parameter h (or σ^{-2}) will be called the precision of the process. So our parameter space consists of all pairs (β, h) with $h > 0$. It follows from (4.2) that the joint density of $(\varepsilon(1), \ldots, \varepsilon(n))$ can be written as

$$(4.3) \qquad f_N^{(n)}(\varepsilon|0, hI) = (2\pi)^{-\frac{1}{2}n} \, e^{-\frac{1}{2}h\varepsilon^t\varepsilon} \, h^{\frac{1}{2}n}$$

where t denotes transposition. From (4.3) it follows that the likelihood of the sample can be written as

$$(4.4) \qquad l(y|\beta, h) = (2\pi)^{-\frac{1}{2}n} \, e^{-\frac{1}{2}h(y-X\beta)^t(y-X\beta)} \, h^{\frac{1}{2}n}$$

and that it has a kernel

$$(4.5) \qquad e^{-\frac{1}{2}h(y-X\beta)^t(y-X\beta)} \, h^{\frac{1}{2}n}$$

In the transition of (4.3) to (4.4) we have assumed that the Jacobian of the transformation $y = X\beta + \varepsilon$ which carries ε into y equals unity. A sufficient condition for this being true is

$$(4.6)\quad \begin{cases} \dfrac{\partial y(t)}{\partial \varepsilon(t)} = 1 & \\[2ex] \dfrac{\partial y(s)}{\partial \varepsilon(t)} = 0 & \text{if } s < t \end{cases}$$

So we need not require that the elements of X are fixed, but we may allow that X has random columns (for instance, lagged values of y; see Chapter 6) if we assume that the disturbance in period t is independent of all past values of the X-variables.

If we let b denote any (or 'the' if rank $[X] = r$) solution of the 'normal equations'

$$(4.7\text{a})\quad X^t Xb = X^t y$$

where b is an r-dimensional vector, and define[1]

$$(4.7\text{b})\quad N = X^t X$$

$$(4.7\text{c})\quad m = \text{rank}\,[X] = \text{rank}\,[N]$$

$$(4.7\text{d})\quad \lambda = n - m$$

$$(4.7\text{e})\quad v = (y - Xb)^t (y - Xb)/\lambda$$

then the kernel (4.5) can be written in the form

$$(4.8)\quad \mathrm{e}^{-\frac{1}{2}h\lambda v}\, h^{\frac{1}{2}\lambda}\, \mathrm{e}^{-\frac{1}{2}h(\beta-b)^t N(\beta-b)}\, h^{\frac{1}{2}m}$$

which is easily verified by writing $\beta = b + (\beta - b)$ and substituting (4.7a)–(4.7e).

In this book we always consider the case that neither β nor h is known. The sample statistic (b, v, N, λ) is then sufficient. The most convenient prior distribution of (β, h) – the natural conjugate of (4.8) – is a Normal-gamma[2] density:

1. Note that $\bar{R}^2$ defined in (1.3.2) can be rewritten as $1 - v/(y^t Ey/n)$; use is made of (4.7a) and (4.7e). Accordingly, given y, there is a monotonically decreasing relationship between v and $\bar{R}^2$.
2. Note that (4.9) contains $h^{\frac{1}{2}\lambda' - 1}$ instead of $h^{\frac{1}{2}\lambda'}$. For an explanation of this we refer to RAIFFA and SCHLAIFER (1961), p. 52. Note the misprint in that page: the complete integral of the kernel exists not only for $\lambda > 0$, $v > 0$ but also for $0 > \lambda > -2$, $v < 0$.

(4.9) $$f_{N\gamma}^{(r)}(\beta, h|b', v', N', \lambda') \equiv$$

$$\equiv \frac{|N'|^{\frac{1}{2}} (\frac{1}{2}\lambda' v')^{\frac{1}{2}\lambda'}}{(2\pi)^{-\frac{1}{2}m'} (\frac{1}{2}\lambda' - 1)!} \mathrm{e}^{-\frac{1}{2}h\lambda' v'} h^{\frac{1}{2}\lambda' - 1} \mathrm{e}^{-\frac{1}{2}h(\beta - b')^t N'(\beta - b')} h^{\frac{1}{2}m'}$$

with $-\infty < \beta < \infty$, $h > 0$, λ', $v' > 0$, and where N' measures the prior information on β in units of the process precision h and the abbreviation m' denotes rank $[N']$. For (4.9) to represent a proper distribution, we must have $m' = r$ and N' to be a positive definite symmetric (PDS) matrix. A kernel of (4.9) is

(4.10) $$\mathrm{e}^{-\frac{1}{2}h\lambda' v'} h^{\frac{1}{2}\lambda' - 1} \mathrm{e}^{-\frac{1}{2}h(\beta - b')^t N'(\beta - b')} h^{\frac{1}{2}m'}$$

With this prior distribution with parameters (b', v', N', λ') and the sample with sufficient statistic (b, v, N, λ), the posterior density of (β, h) can be obtained.[1] It is Normal-gamma with parameters

(4.11 a) $N'' = N' + N = N' + X^t X$

(4.11 b) $b'' = (N'')^{-1} (N' b' + Nb)$

(4.11 c) $m'' = \text{rank}\,[N'']$

(4.11 d) $\lambda'' = [\lambda' + m'] + [\lambda + m] - m''$

(4.11 e) $v'' = [(\lambda' v' + b'^t N' b') + (\lambda v + b^t Nb) - b''^t N'' b'']/\lambda''$

These results apply regardless of the rank of X.

If the joint posterior distribution of (β, h) is Normal-gamma, then it is possible to show that the marginal distribution of h is gamma-2 with density

(4.12) $$D(h|b'', v'', N'', \lambda'') = f_{\gamma 2}(h|v'', \lambda'') \propto$$
$$\propto \mathrm{e}^{-\frac{1}{2}h\lambda'' v''} h^{\frac{1}{2}\lambda'' - 1}$$

Since the parameter N'' of this joint distribution is of full rank,[2] it is possible to show that the marginal distribution of β is Student with λ'' degrees of freedom and with density

(4.13) $$D(\beta|b'', v'', N'', \lambda'') = f_S^{(r)}(\beta|b'', N''/v'', \lambda'') \propto$$
$$\propto [\lambda'' v'' + (\beta - b'')^t N''(\beta - b'')]^{-\frac{1}{2}(m'' + \lambda'')}$$

1. For proofs of the results (4.11)–(4.15), see RAIFFA and SCHLAIFER (1961), Section 13.5. For more details about gamma-2 distributions, see, e.g., RAIFFA and SCHLAIFER (1961), Section 7.6. For more details on (multivariate) Student distributions, see, e.g., RAIFFA and SCHLAIFER (1961), Section 8.3.
2. This can easily be seen if one recalls that N' has full rank.

and, therefore with mean and variance

(4.14) $\quad E(\beta|b'', v'', N'', \lambda'') = b'' \qquad$ if $\lambda'' > 1$

(4.15) $\quad V(\beta|b'', v'', N'', \lambda'') = (N'')^{-1}(\lambda'' v''/(\lambda'' - 2)) \qquad$ if $\lambda'' > 2$

The conditional distribution of β given h is, of course, normal with density $f_N^{(r)}(\beta|b'', hN'')$.

2.5. SOME REMARKS ABOUT THE NORMAL-GAMMA PRIOR DISTRIBUTION

We start this section to discuss how to specify the parameters of the Normal-gamma prior distribution as presented in (4.8). In this we mainly follow Harkema and Kloek (1969). Theil (1963) points out that it will be extremely difficult in many (econometric) applications to make sensible *a priori* statements about the magnitude of σ^2 by means of the procedure suggested by Raiffa and Schlaifer (1961). As eventual prior knowledge (in our opinion) bears more directly on the size of the disturbances than on the size of their standard deviation σ, we believe that it is more easy to specify the prior information on the basis of the unconditional distribution of the disturbances, rather than by means of the inverted-gamma-2 distribution of σ that results from the gamma-2 marginal prior distribution of h. Since we have assumed that, given h, the disturbances $\varepsilon(t)$ $(t = 1, \ldots, n)$ are identically and independently distributed normal variables with zero mean and unknown precision h, we obtain

(5.1) $\quad l(\varepsilon(1), \ldots, \varepsilon(n)|h) = (2\pi)^{-\frac{1}{2}n} h^{\frac{1}{2}n} \exp\{-\tfrac{1}{2}h\varepsilon^t\varepsilon\}$

Upon multiplying (5.1) and the marginal prior distribution of h we find that the joint distribution of (ε, h) is Normal-gamma, and integrating this distribution with respect to h one finds that the marginal distribution of ε is Student with λ' degrees of freedom and

(5.2) $\quad E(\varepsilon|\lambda', v') = 0 \qquad$ if $\lambda' > 1$

$\quad V(\varepsilon|\lambda', v') = (\lambda' v'/(\lambda' - 2))I \qquad$ if $\lambda' > 2$

From (4.15) it is easy to derive that the marginal prior distribution of β is also Student with λ' degrees of freedom and with variance matrix $(N')^{-1}(\lambda' v'/(\lambda' - 2))$ if $\lambda' > 2$. Note that the same λ' and v' occur in both the marginal distribution of ε and the marginal prior distribution of β, and hence in the marginal prior distribution of every element of β. This property causes some restrictions on the possibilities to express prior knowledge within the family of Normal-gamma prior distributions.

For the specification of the values b', v', N', and λ', some probability statements based on 'subjective betting odds' have to be made. It is not easy to give general rules. An illustrative and amusing example how to tackle the problem of assigning probability statements based on the subjective betting odds[1] can be found in Raiffa (1968), pp. 161–168. He first specifies the median, then the odd quartiles, and finally the odd octiles. In our case, we have assumed that the median of the disturbances is equal to zero. Further, the symmetric marginal distributions of β and ε imply the following restrictions: (i) the first quartile and the third quartile have equal distances to the median; (ii) the first octile and the seventh octile also have equal distances to the median; (iii) and finally, the third octile and the fifth octile have equal distances to the median. Now, two procedures seem available to obtain the parameters b', v', N', and λ';

(i) A simultaneous approach. The investigator specifies simultaneously the medians, odd quartiles, and odd octiles of β and ε. Independent specification of these values will often give rise to contradictions in finding an appropriate prior Normal-gamma distribution, unless the odd octiles are neglected. If not, these contradictions must be reconciled by means of an adjustment procedure. (ii) A stepwise approach. Then first the information is formulated on ε and next on β, or the other way round.

We continue to discuss in more detail the stepwise approach which starts with the disturbances. As mentioned before, the median of the unconditional distribution of ε is equal to zero. Next, it seems quite reasonable that one presumes that half of the disturbances lay in a certain interval which is symmetric around the (zero) median, viz.,

(5.3) $$P[-\varepsilon^1 < \varepsilon < \varepsilon^1] = 0.50$$

where ε^1 is some positive real number. This corresponds with assigning the upper and lower quartiles of the unconditional distribution of ε. Finally, a conditional probability statement is needed, e.g., on the median of ε given that $\varepsilon > \varepsilon^1$. This yields

(5.4) $$P[\varepsilon^1 < \varepsilon < \varepsilon^2 | \varepsilon > \varepsilon^1] = 0.50$$

1. Note that actual betting on β and h is impossible since the 'true' values of β and h will never be known. However, betting on β and h may help us in betting on y and since y is observable a pay-off function may be attached to it. We refer to Pratt, Raiffa, and Schlaifer (1964) who state: 'In evaluating *hypothetical* lotteries whose payoffs depend in whole or in part on x and/or y, we shall ask the decision maker to imagine that he will *never actually learn* the true values of these quantities – what he will learn is simply the consequence that the lottery pays out.' Here (x, y) denote the outcome of an imaginary canonical experiment.

where ε^2 denotes some positive real number such that $\varepsilon^2 > \varepsilon^1$. Combining (5.3) and (5.4) we obtain

$$(5.5) \qquad P[-\varepsilon^2 < \varepsilon < \varepsilon^2] = 0.75$$

Now it is possible to derive values of λ' and v' so that the fit of the Student density of ε to the probability statements (5.3) and (5.5) is optimized. This can be done with the aid of a table of cumulative Student distributions. Exact fitting would be a coincidence, since the existing tables only contain integer values of λ'. If the investigator is able and willing to specify more conditional probability statements of the type (5.4) on the disturbances, and if he has at his disposal[1] the needed statistical tables containing the necessary data for the corresponding unconditional probability statements, then a minimizing procedure could be developed such that the Student density of the disturbances optimally fits all these probability statements.

Now, the investigator still has to specify his prior ideas about the values of the parameter vector β. He can start with assigning a 50 per cent confidence interval to every element of β. Then b' is determined. Given the fact that λ' and v' have already been chosen, the matrix N' is then implicitly defined if one assumes that all elements of β are uncorrelated. If not, some (or all) off-diagonal elements of N' still have to be specified.

The other stepwise approach to obtain the parameters b', v', N', and λ' is to start with formulating the prior ideas about β. This approach enables the investigator to formulate more probability statements on β than in the approach discussed above, since λ' and v' are still free to choose. Note, however, that every element of β has a marginal density which is Student with the same λ'. Following this way the investigator obtains values for b', $v'N'$, and λ'. One probability statement on the unconditionally distributed disturbances, finally, yields the value of v', so that N' also comes available.

In the foregoing some suggestions have been made how to specify the values b', v', N', and λ' of the Normal-gamma prior distribution. On the one hand such a natural-conjugate prior distribution is analytically tractable (see Section 2.3) and enables to incorporate prior information, however scarce, which can have great importance in the case of a small sample (compare Section 2.6). On the other hand, we believe that it is a rather hard job to specify the values b', v', N', and λ' in a given practical case. To

1. The well-known statistical tables of PEARSON and HARTLEY (1956) and FISHER and YATES (1957) do not contain cumulative Student tables for non-integer values of λ'. Moreover, these tables do not contain the data corresponding to the odd octiles mentioned before. For integer values of λ', however, LEMPERS and LOUTER (1971) extended the table of the Student distribution to the odd octiles and odd sediciles (fractiles where the distribution function takes values which are multiples of 1/16).

explain this statement, some critical comments are now in order about the disadvantages of this prior distribution. First, the necessity to specify something about the order of magnitude of the disturbances. If some historical evidence is available this might be rather simple, but in the case one is confined to subjective betting odds, we presume that most people will dislike this job. Secondly, the marginal distribution of β is symmetric. In many cases the investigator can be convinced that his prior ideas do not coincide with such a symmetric framework. Other prior distributions then can be used, such as Beta distributions, although analytical results can not be obtained and numerical integration methods have to be applied. Another possible solution to this problem is to use a weighted sum of Student distributions, such that the non-symmetric prior ideas are formulated within the framework of natural-conjugate prior distributions. This approach, however, seems to be very time-consuming, since such a combination of distributions has to be found by trial and error. Up till now, no analytical tool to obtain such a prior has been developed. Third, β and ε both have a marginal Student distribution with the same λ'. This explains the preceding differences in starting with β or ε. If one starts with one of them, only one probability statement still can be made about the other. This is an unattractive restriction. A final disadvantage of the Normal-gamma prior distribution is the implicitly defined dependence of β and h. This property can easily be seen by the fact that the marginal distribution of β is Student, while, on the other hand, the conditional distribution of β given h is normal. For increasing values of λ', however, the Student distribution converges to the normal distribution. The dependence of β and h implies, that if the investigator would be able to specify a marginal prior distribution on β, and dislike to say something about the disturbances, the Normal-gamma approach is unable to handle such a situation. Leamer (1970), however, points out that for large samples an independent Normal-gamma prior can be used. Then β is normally distributed independent of the precision h, and the gamma-2 distribution on h is retained. The posterior distribution is integrated out over h to obtain the marginal posterior distribution of β which is proportional to the product of the normal prior of β and a Student density of y given β. This Student density can be rewritten as a Student function in β and, as the posterior degree of freedom grows, can be approximated by a Normal distribution. As a consequence, the marginal posterior distribution of β becomes Normal.

2.6. UNIFORM PRIOR DISTRIBUTIONS

In case the investigator knows very little about his parameters he often uses a uniform prior distribution. This prior distribution provides a way of expressing ignorance of the value(s) of the parameter(s), and was suggested by Jeffreys (1961) and even by Savage (1962). It also provides a way to attain that the sample information dominates in the posterior distribution. It reads as follows: the prior distributions of β and $\log \sigma$ should be taken as locally independent and uniform.[1] That is, if we substitute $h = \sigma^{-2}$,

$$(6.1) \qquad D'(\beta) \propto k_1$$

$$(6.2) \qquad D'(h) \propto h^{-1}$$

with $-\infty < \beta < \infty$, $h > 0$, and k_1 some arbitrary positive constant. On combining the prior distributions (6.1) and (6.2) with the likelihood function (4.8) by means of Bayes' Theorem, we obtain the following Normal-gamma posterior density of β and h

$$(6.3) \qquad D''(\beta, h|y) \propto h^{\frac{1}{2}n-1} \exp[-\tfrac{1}{2}h\{\lambda v + (\beta-b)^t N(\beta-b)\}]$$

which is a proper density provided that N has full rank and λ and v are positive. The marginal posterior density of β is obtained by integrating (6.3) over h, to yield

$$(6.4) \qquad D''(\beta|y) \propto [\lambda v + (\beta-b)^t N(\beta-b)]^{-\frac{1}{2}n}$$

which is of course a multivariate Student density. It can easily be shown that the marginal posterior distribution of a subset of the elements of β is also of the same form. In particular, the marginal distribution of a single element of β can be expressed in terms of a univariate Student distribution with λ degrees of freedom.

Some remarks have to be made. Using the uniform prior distributions means that the investigator is very careless with his prior information. The advantage of this approach is that it is rather easy to apply, and for large samples it is superfluous to specify a more sophisticated prior distribution. On the other hand, we already mentioned in Section 2.3 that the concept of 'no information' is not well-defined.[2] A consequence of the uniform-

1. Of course these distributions are no proper distributions since the densities have no finite integrals. But if we substitute them in (2.1), there are no integrability problems with (2.2). A more detailed discussion can be found in Section 3.5, where we demonstrate that the same results can be reached by considering a sequence of uniform distributions on finite intervals.
2. For convenience, however, we shall make use of the term 'no prior information' if we mean vague or little prior knowledge.

prior-distribution approach is, e.g., that $P[-100 < \beta < 100] = 0$, where β is, for example, an income elasticity. This, of course, is absurd. Finally, if one has some information, however scarce, this information can have great importance in the case of small samples. In such a case, it would be wasteful to apply the uniform prior distributions.

2.7. SEQUENCES OF NORMAL-GAMMA DENSITIES WITH DECREASING QUANTITIES OF PRIOR INFORMATION

It has been suggested more or less implicitly by Raiffa and Schlaifer (1961) that uniform prior densities can be considered as limiting cases of natural-conjugate densities. In the present section we show that this is not generally true and point out the remarkable difference.[1]

Let $\{D'_\alpha(\beta, h)\}_{\alpha=1}^{\infty}$ be a sequence of Normal-gamma prior densities with parameters b'_α, v'_α, N'_α, and λ'_α, as defined in (4.9), apart from the α's. We require (for $\alpha = 1, 2, \ldots$) that

(7.1) $\qquad v'_\alpha > 0 \qquad \lambda'_\alpha > 0 \qquad N'_\alpha$ is PDS

in order to guarantee that all densities are proper. In addition we require that the sequences $\{b'_\alpha\}_{\alpha=1}^{\infty}$ and $\{v'_\alpha\}_{\alpha=1}^{\infty}$ are bounded. Note that (7.1) implies

(7.2) $\qquad m'_\alpha = \text{rank}[N'_\alpha] = r \qquad\qquad (\alpha = 1, 2, \ldots)$

(7.3) $\qquad \lim_{\alpha\to\infty} m'_\alpha = r$

We shall describe the decreasing quantity of information by supposing[2]

(7.4) $\qquad \lim_{\alpha\to\infty} \lambda'_\alpha = 0 \qquad \lim_{\alpha\to\infty} N'_\alpha = 0$

It follows that we encounter the following property

(7.5) $\qquad \lim_{\alpha\to\infty} \{\text{rank}[N'_\alpha]\} \neq \text{rank}[\lim_{\alpha\to\infty} N'_\alpha]$

Next, we consider a sample of n observations as discussed in Section 2.4, which is sufficiently described by the statistic (b, v, N, λ), and assume

(7.6) $\qquad m \equiv \text{rank}\,[N] = r \qquad v > 0$

1. We largely follow LEMPERS, HARKEMA, and KLOEK (1969) with respect to the mathematics, not with respect to their comments.
2. Compare RAIFFA and SCHLAIFER (1961), page 63, where they discuss the case that $N' \to 0$ for a simple case. For λ' an analogous reasoning can be given.

which implies

(7.7) $\quad n \geqslant r+1$

Combining the sequence of prior densities and the likelihood of this sample, we obtain a sequence of Normal-gamma posterior densities $\{D''_\alpha(\beta, h)\}_{\alpha=1}^{\infty}$, with parameters

(7.8a) $\quad N''_\alpha = N'_\alpha + N$

(7.8b) $\quad b''_\alpha = (N''_\alpha)^{-1}(N'_\alpha b'_\alpha + Nb)$

(7.8c) $\quad m''_\alpha = \text{rank}[N''_\alpha]$

(7.8d) $\quad \lambda''_\alpha = [\lambda'_\alpha + m'_\alpha] + [\lambda + m] - m''_\alpha$

and

(7.8e) $\quad v''_\alpha = [\lambda'_\alpha v'_\alpha + b'^t_\alpha N'_\alpha b'_\alpha + \lambda v + b^t Nb - b''^t_\alpha N''_\alpha b''_\alpha]/\lambda''_\alpha$

Raiffa and Schlaifer (1961), p. 319 state (without giving an argument for their statement) that $N' = 0$ in the case of no information. If, however, one considers no information as a limiting case of decreasing information one obtains a different result. To show this we make use of the limit results (7.3) and (7.4). Taking limits of the sequences of parameters we find

(7.9a) $\quad \lim\limits_{\alpha\to\infty} N''_\alpha = N$

(7.9b) $\quad \lim\limits_{\alpha\to\infty} b''_\alpha = b$

(7.9c) $\quad \lim\limits_{\alpha\to\infty} m''_\alpha = \lim\limits_{\alpha\to\infty} \text{rank}\,[N''_\alpha] = r$

(7.9d) $\quad \lim\limits_{\alpha\to\infty} \lambda''_\alpha = \lambda + m = n$

and

(7.9e) $\quad \lim\limits_{\alpha\to\infty} v''_\alpha = \lambda v/n$

We proceed to consider the density

(7.10) $\quad D''(\beta, h) \equiv f_{N\gamma}^{(r)}(\beta, h|b, \lambda v/n, N, n) =$

$$= \frac{|N|^{\frac{1}{2}}(\frac{1}{2}\lambda v)^{\frac{1}{2}n}}{(2\pi)^{\frac{1}{2}r}(\frac{1}{2}n-1)!} h^{\frac{1}{2}n+\frac{1}{2}r-1} \exp[-\tfrac{1}{2}h\{\lambda v + (\beta - b)^t N(\beta - b)\}]$$

which is obtained by replacing the parameters of $D''_\alpha(\beta, h)$ by their limits. It follows from assumption (7.6) that (7.10) is a proper density. Therefore, we may apply Scheffé's Theorem[1] and conclude that the sequence $\{D''_\alpha(\beta, h)\}_{\alpha=1}^{\infty}$ converges in distribution to $D''(\beta, h)$.

The suggestion to accept $N'=0$ with rank $[N']=0$, however, would have resulted in the posterior distribution of (β, h) presented in (6.3). So an identical result is obtained with that of the approach with uniform prior densities. The difference between (6.3) and (7.10) is the power of h. As a straightforward consequence of (7.10), the marginal distribution of β is Student with parameters $(b, nN/\lambda v, n)$ and the marginal distribution of h is gamma-2 with parameters $(\lambda v/n, n)$. It is derived in Section 2.6 that in the case of uniform prior distributions, the marginal distribution of β is Student with parameters $(b, N/v, \lambda)$. For completeness we mention that in that case the marginal distribution of h is gamma-2 with parameters (v, λ).

It is interesting to discuss in some details the differences of the two alternative approaches. Lempers, Harkema and Kloek (1969) conclude that the lengths of arbitrary central confidence intervals of the marginal distribution of a β element are smaller for the limit-Normal-gamma case than for the uniform case. These differences are, as they pointed out, considerable for small samples. This is illustrated in Table 2.1 below for 95 per cent Bayesian confidence intervals. The percentage reductions in this table are defined as the difference of the lengths of 95 per cent Bayesian confidence intervals in the uniform case and the limit-Normal-gamma case, divided by the length in the uniform case, and multiplied by 100. The results demonstrate the danger of specifying vague prior distributions if one has a small sample.

Another difference between the two cases of no prior information is that β and h are independent in the uniform approach and dependent in the Normal-gamma approach. A similar type of dependence also arises in the invariance-theory approach discussed by Jeffreys (1961), Chapter III, § 3.10. Jeffreys rejects this approach because he believes that the investigator's prior information on h will always be independent of that on β.

Let us investigate in more detail the dependence of β and h in the limit-Normal-gamma approach. The α-th conditional prior distribution of β, given h, is normal with parameters (b'_α, hN'_α). It does not converge to a proper distribution as α tends to infinity, but it becomes increasingly uniform in the sense that the ratio of the densities at two arbitrary distinct points β_1 and β_2 tends to unity. The proof is a straightforward generalization of the proof given by Raiffa and Schlaifer (1961), p. 296. The marginal distribution of β is Student with parameters $(b'_\alpha, N'_\alpha/v'_\alpha, \lambda'_\alpha)$. Its limiting

1. See Scheffé (1947).

TABLE 2.1. PERCENTAGE REDUCTION OF 95 PER CENT BAYESIAN CONFIDENCE INTERVALS*

Sample size $\lambda+m=n$	*Number of Explanatory Variables* $(r=m)$ *1*	*2*	*3*	*4*	*5*	*6*	*7*	*8*	*9*	*10*
2	76.05	—	—	—	—	—	—	—	—	—
3	39.62	85.54	—	—	—	—	—	—	—	—
4	24.25	54.38	89.08	—	—	—	—	—	—	—
5	17.16	37.41	62.21	90.95	—	—	—	—	—	—
6	13.12	28.03	45.62	67.17	92.14	—	—	—	—	—
7	10.52	22.26	35.60	51.34	70.62	92.96	—	—	—	—
8	8.79	18.39	29.09	41.26	55.62	73.20	93.58	—	—	—
9	7.52	15.65	24.52	34.42	45.68	58.96	75.22	94.07	—	—
10	6.56	13.58	21.18	29.47	38.72	49.24	61.65	76.84	94.45	—
11	5.81	11.99	18.60	25.76	33.57	42.28	52.19	63.88	78.19	94.78
12	5.21	10.72	16.57	22.85	29.63	37.03	45.29	54.68	65.76	79.33
13	4.76	9.73	14.97	20.55	26.52	32.98	40.03	47.90	56.84	67.39
14	4.31	8.86	13.61	18.63	23.97	29.68	35.87	42.61	50.14	58.70
15	4.02	8.15	12.53	17.09	21.91	27.03	32.51	38.45	44.92	52.15
16	3.68	7.55	11.53	15.74	20.14	24.78	29.71	34.99	40.71	46.95
17	3.44	6.99	10.73	14.58	18.64	22.89	27.37	32.13	37.23	42.75
18	3.23	6.56	10.00	13.62	17.34	21.27	25.38	29.71	34.32	39.26
19	3.04	6.17	9.40	12.73	16.24	19.85	23.66	27.64	31.85	36.32
20	2.86	5.81	8.85	12.99	15.23	18.64	22.14	25.85	29.71	33.80
21	2.69	5.47	8.34	11.31	14.36	17.51	20.82	24.23	27.84	31.60
22	2.58	5.20	7.91	10.71	13.59	16.57	19.64	22.87	26.19	29.70
23	2.43	4.95	7.51	10.15	12.88	15.70	18.60	21.59	24.75	27.99
24	2.34	4.72	7.18	9.68	12.26	14.92	17.67	20.51	23.43	26.51
25	2.21	4.50	6.82	9.23	11.67	14.20	16.80	19.49	22.26	25.12
26	2.13	4.30	6.54	8.81	11.17	13.56	16.03	18.58	21.21	23.92
27	2.06	4.15	6.27	8.46	10.69	13.00	15.34	17.76	20.25	22.83
28	1.99	4.01	6.06	8.14	10.29	12.47	14.73	17.02	19.40	21.84
29	1.88	3.84	5.82	7.83	9.87	11.98	14.12	16.34	18.59	20.91
30	1.83	3.67	5.59	7.54	9.51	11.51	13.58	15.69	17.86	20.07

* Due to the limited precision of the existing tables of percentage points of the Student distribution, the last decimal place is afflicted with rounding errors.

behavior is not uniquely defined by (7.4); it depends on the limiting behavior of $N'_\alpha/\lambda'_\alpha$. If

$$\lim_{\alpha\to\infty} \frac{N'_\alpha}{\lambda'_\alpha} = 0 \tag{7.11}$$

the ratio of the densities at two arbitrary points β_1 and β_2 tends to unity (which can be interpreted as uniformity); otherwise the limit depends on β_1 and β_2. These results are readily obtained by generalizing the argument of Raiffa and Schlaifer (1961), pp. 302–303.

If we apply the same limit-ratio procedure to the conditional density of h given β we find $(h_1/h_2)^{\frac{1}{2}r-1}$; if we apply it to the marginal distribution of h, we find h_2/h_1. Use has been made of the formulas (7–50), (13–32), and (13–33) of Raiffa and Schlaifer (1961). This demonstrates that the dependence of β and h does not vanish in the limit.

Although Jeffreys rejects the dependence of β and h, as already stated before, we would, however, suggest that an investigator whose prior information is vague at least should ask himself the question whether β and h in his prior ideas are dependent or not. Accordingly, he then can adopt one approach or the other. Geisel (1970), De Groot (1970), and Zellner (1971) use the limit-Normal-gamma approach. This approach also can be used to be 'informative' on some elements of β and to be 'noninformative' on the other elements, without loosing the attractive property of tractability. For a more detailed discussion of this possibility we refer to Harkema (1971), Chapter 3.

Up till now, however, the uniform prior distributions play a more important role in statistical studies in the case of no prior information. We refer to, e.g., Zellner and Tiao (1964), Tiao and Zellner (1964), Zellner and Chetty (1965) and Zellner and Geisel (1968). Therefore, in the remainder of this book we make use of such uniform prior distributions in the case of vague or no prior information. In some cases, however, we compare the results obtained with the two alternative approaches. For the case that prior information is available, we confine ourselves in this book to the natural-conjugate approach.

Finally, we note that for the uniform as well as for the limit natural-conjugate approach it is assumed that N has full rank. This assumption is necessary in order to obtain proper posterior distributions. For the natural-conjugate approach, however, this assumption is superfluous, since N'' has full rank; see Section 2.4.

3. Posterior probabilities of alternative linear models

3.1. INTRODUCTION

In Chapter 2 the necessary tools were presented which will enable us in this chapter to develop the posterior probabilities of alternative specifications of linear models. The results as presented here are mainly based on a paper by Kloek and Lempers (1970). We derive a mathematical framework, while most of the applications are discussed in the Chapters 4, 5, and 6.

In this chapter the order of discussion is as follows. In Section 3.2 the general posterior probability of a hypothesis out of a set of mutually exclusive hypotheses is derived; Sections 3.3, 3.4, and 3.5 apply this result to a set of alternative linear hypotheses for natural-conjugate as well as uniform prior distributions. In Section 3.6, a large sample property is derived, and in Section 3.7 some preliminary comments are made about applicability. Finally, in Section 3.8, a simple particular case is discussed. Appendix A contains a rather technical proof for a result required in Section 3.6.

3.2. POSTERIOR PROBABILITIES OF SUBSETS OF THE STATE SPACE

Let us consider a state space Θ which can be partitioned into a finite[1] number of measurable subspaces $\Theta_1, \ldots, \Theta_k$ as follows:

(2.1) $$\Theta = \bigcup_{i=1}^{k} \Theta_i$$

(2.2) $$\Theta_i \cap \Theta_j = \phi \qquad (i, j = 1, \ldots, k; \quad i \neq j)$$

We denote the hypothesis that $\theta \in \Theta_i$ by H_i and the probability that H_i is true by p_i so that

(2.3) $$p_i \equiv P[\theta \in \Theta_i]$$

1. The generalization to a countable number is straightforward but not relevant for the applications. In particular cases even uncountable numbers can be handled. See Section 4.5.

We consider the case that all Θ_i are uncountable and that for each i we have a conditional prior density $D'(\theta|\Theta_i)$. Let $K'(\theta|\Theta_i)$ be a kernel of $D'(\theta|\Theta_i)$; that is, let

(2.4) $$D'(\theta|\Theta_i) = C_i' K'(\theta|\Theta_i)$$

where C_i' is implicitly defined by

(2.5) $$C_i' \int_{\Theta_i} K'(\theta|\Theta_i)\,\mathrm{d}\theta = 1$$

If the condition $\theta \in \Theta_i$ is dropped, the prior[1] is $p_i' D'(\theta|\Theta_i)$ $(\theta \in \Theta_i;\ i = 1, \ldots, k)$. Let $l_i(z|\theta)$ denote the likelihood of the sample $z = (z(1), \ldots, z(n))$ given θ $(\theta \in \Theta_i)$. This likelihood can be written as

(2.6) $$l_i(z|\theta) = \kappa_i(z|\theta)\,\rho_i(z)$$

where $\kappa_i(z|\theta)$ is a kernel of the likelihood of z given θ $(\theta \in \Theta_i)$. Then the posterior density of $\theta \in \Theta_i$, given z, reads

(2.7) $$D''(\theta|z) = \frac{p_i' D'(\theta|\Theta_i)\, l_i(z|\theta)}{\sum_j p_j' \int_{\Theta_j} D'(t|\Theta_j) l_j(z|t)\mathrm{d}t} = \frac{p_i' C_i' \rho_i(z)\, K'(\theta|\Theta_i)\, \kappa_i(z|\theta)}{\sum_j p_j' C_j' \rho_j(z) \int_{\Theta_j} K'(t|\Theta_j)\kappa_j(z|t)\mathrm{d}t} \qquad (\theta \in \Theta_i;\ i = 1, \ldots, k)$$

and the posterior probability that H_i is true

(2.8) $$p_i'' \equiv P''[\Theta_i|z] = \int_{\Theta_i} D''(\theta|z)\,\mathrm{d}\theta = \frac{p_i' C_i' \rho_i(z) \int_{\Theta_i} K'(t|\Theta_i)\,\kappa_i(z|t)\,\mathrm{d}t}{\sum_j p_j' C_j' \rho_j(z) \int_{\Theta_j} K'(t|\Theta_j)\,\kappa_j(z|t)\,\mathrm{d}t}$$

Later we shall discuss the ratio of the posterior probabilities of two alternative hypotheses

(2.9) $$\frac{p_i''}{p_j''} = \frac{p_i' \rho_i(z)\, C_i'/C_i''(z)}{p_j' \rho_j(z)\, C_j'/C_j''(z)}$$

1. Note that this prior is a mixed mass-density function.

where $C_i''(z)$ is defined by

(2.10) $$C_i''(z)\int_{\Theta_i} K'(t|\Theta_i)\,\kappa_i(z|t)\,\mathrm{d}t = 1$$

From (2.9) we conclude that the ratio of the integration constants C_i' and C_j' of the prior distributions plays a crucial role in the analysis.

3.3. THE CASE OF SEVERAL LINEAR MODELS: THE LIKELIHOOD FUNCTION

In comparing different models to explain one variable, one usually uses the same sample z. The formulation of these models, however, may have been chosen in such a way that the variables to be explained are different; in such a case, they are in general functionally related. Therefore, we introduce the linear model as $y_i = X_i\beta_i + \varepsilon_i$, where $y_i = f_i(z)$ and f_i is a known bijective mapping which sends the sample space $Z = \{z\} \subset R^n$ into R^n; well-known examples are $y_i(t) = \log z(t)$ $(t = 1, \ldots, n)$ and $y_i = A_i z$ where A_i is a known non-singular matrix. Furthermore[1] X_i is a matrix of order $n \times r_i$ and rank m_i, and ε_i is a normally distributed random n-vector of disturbances with $E(\varepsilon_i) = 0$ and $E(\varepsilon_i\varepsilon_i^t) = (1/h_i)I$, where h_i denotes the precision of the i-th process, that is, the reciprocal of its variance. Finally, we need not require that X_i is fixed, but assume $\partial y_i(t)/\partial\varepsilon_i(t) = 1$ and $\partial y_i(s)/\partial\varepsilon_i(t) = 0$ if $s < t$ (compare (2.4.6)). Given these assumptions, the likelihood of the sample is

(3.1) $$l_i(z|\theta) = l(y_i|\theta)\,|J_i| = c_i \mathrm{e}^{-q_i}\, h_i^{s_i}|J_i| \qquad (\theta \in \Theta_i)$$

with

(3.2a) $$c_i = (2\pi)^{-\frac{1}{2}n}$$

(3.2b) $$q_i = \tfrac{1}{2}h_i\{(\beta_i - b_i)^t N_i(\beta_i - b_i) + \lambda_i v_i\}$$

(3.2c) $$s_i = \tfrac{1}{2}m_i + \tfrac{1}{2}\lambda_i$$

(3.2d) $$|J_i| = |\det[\partial y_i(s)/\partial z(t)]|$$

and

(3.2e) $$\rho_i(z) = c_i|J_i| = (2\pi)^{-\frac{1}{2}n}|J_i|$$

1. It may happen that X_i is a submatrix of X_j, which implies $\Theta_i \subset \Theta_j$ contrary to (2.2). This problem can easily be solved by defining a new subspace $\Theta_j^* = \Theta_j - \Theta_i$. Since Θ_j^* contains 'almost every' point of Θ_j and since we confine ourselves to continuous distributions on Θ_j, corresponding integrals over Θ_j and Θ_j^* are equal.

where

(3.3a) $N_i = X_i^t X_i$

(3.3b) $(X_i^t X_i) b_i = X_i^t y_i$

(3.3c) $m_i = \text{rank}[N_i]$

(3.3d) $\lambda_i = n - m_i$

(3.3e) $v_i = (y_i - X_i b_i)^t (y_i - X_i b_i)/\lambda_i$

see Section 2.4, or Raiffa and Schlaifer (1961), Chapter 13.

3.4. PRIOR AND POSTERIOR ANALYSIS WITH NATURAL-CONJUGATE PRIOR DISTRIBUTIONS

In specifying a conditional Natural-conjugate prior density for the parameters of the likelihood function (3.1), we suppose that this prior distribution has been fitted to subjective betting odds.[1] We follow Raiffa and Schlaifer (1961), Chapter 13, and take the Normal-gamma density

(4.1) $D'(\beta_i, h_i|\Theta_i) = C_i' \, e^{-q_i'} h_i^{s_i'}$

with

(4.2a) $C_i' = (2\pi)^{-\frac{1}{2}m_i'} |N_i'|^{\frac{1}{2}} (\frac{1}{2}\lambda_i' v_i')^{\frac{1}{2}\lambda_i'} / (\frac{1}{2}\lambda_i' - 1)!$

(4.2b) $q_i' = \frac{1}{2}h_i\{(\beta_i - b_i')^t N_i' (\beta_i - b_i') + \lambda_i' v_i'\}$

(4.2c) $s_i' = \frac{1}{2}m_i' + \frac{1}{2}\lambda_i' - 1$

where $m_i' = \text{rank}\,[N_i']$. In order to guarantee that (4.1) is a proper density, we assume for $i = 1, \ldots, k$ that $\lambda_i' > 0$, $v_i' > 0$, and N_i' is positive-definite symmetric, so that

(4.3) $m_i' = r_i$

Combining this prior density with the likelihood function (3.1), we find a Normal-gamma posterior density

(4.4) $D''(\beta_i, h_i|z) = C_i'' \, e^{-q_i''} h_i^{s_i''}$

1. One may remark that betting on the question as to which of two models is the 'true' one is difficult since one will never know with certainty whether a model is 'true' or not. A possible solution to this problem might be that the player who bets that model i is 'true' wins if model i yields better conditional predictions than the other model in a previously specified prediction problem. See also Section 1.4.

with

$$(4.5a)\quad C_i'' = (2\pi)^{-\frac{1}{2}m_i''}|N_i''|^{\frac{1}{2}}(\tfrac{1}{2}\lambda_i''v_i'')^{\frac{1}{2}\lambda_i''}/(\tfrac{1}{2}\lambda_i''-1)!$$

$$(4.5b)\quad q_i'' = \tfrac{1}{2}h_i\{(\beta_i-b_i'')^t N_i''(\beta_i-b_i'')+\lambda_i''v_i''\}$$

$$(4.5c)\quad s_i'' = \tfrac{1}{2}m_i''+\tfrac{1}{2}\lambda_i''-1$$

where

$$(4.6a)\quad N_i'' = N_i'+N_i$$

$$(4.6b)\quad b_i'' = (N_i'')^{-1}(N_i'b_i'+N_ib_i)$$

$$(4.6c)\quad m_i'' = \text{rank}[N_i'']$$

$$(4.6d)\quad \lambda_i'' = \lambda_i'+m_i'+\lambda_i+m_i-m_i''$$

and[1]

$$(4.6e)\quad v_i'' = \{\lambda_i'v_i'+b_i'^tN_i'b_i'+\lambda_iv_i+b_i^tN_ib_i-b_i''^tN_i''b_i''\}/\lambda_i''$$
$$= \{\lambda_i'v_i'+\lambda_iv_i+(b_i-b_i')^t\{(N_i')^{-1}+N_i^{-1}\}^{-1}(b_i-b_i')\}/\lambda_i''$$

and find that the ratio of posterior probabilities (2.9) can now be rewritten as

$$(4.7)\quad \frac{p_i''}{p_j''} = \frac{p_i'|N_i'|^{\frac{1}{2}}|N_j''|^{\frac{1}{2}}(\tfrac{1}{2}\lambda_i'v_i')^{\frac{1}{2}\lambda_i'}(\tfrac{1}{2}\lambda_j''v_j'')^{\frac{1}{2}\lambda_j''}}{p_j'|N_j'|^{\frac{1}{2}}|N_i''|^{\frac{1}{2}}(\tfrac{1}{2}\lambda_j'v_j')^{\frac{1}{2}\lambda_j'}(\tfrac{1}{2}\lambda_i''v_i'')^{\frac{1}{2}\lambda_i''}} \times$$
$$\times \frac{|J_i|(\tfrac{1}{2}\lambda_j'-1)!(\tfrac{1}{2}\lambda_i''-1)!}{|J_j|(\tfrac{1}{2}\lambda_i'-1)!(\tfrac{1}{2}\lambda_j''-1)!}$$

3.5. PRIOR AND POSTERIOR ANALYSIS WITH UNIFORM PRIOR DISTRIBUTIONS

In this section we investigate the results of Section 3.2 for the likelihood function of Section 3.3 combined with uniform prior distributions for β and $\log h$. We begin by studying the conditional posterior distributions (for $i=1, \dots, k$) of β_i and h_i (given that the i-th model is the correct one) and drop for convenience the subscript i. Let R^r stand for r-dimensional Euclidean space, R^+ denote the set of all positive real numbers, $\{\beta_{l\alpha}\}_{\alpha=1}^{\infty}$, $\{\beta_{u\alpha}\}_{\alpha=1}^{\infty}$ denote sequences in R^r, where $\beta_{l\alpha} < \beta_{u\alpha}$ ($\alpha = 1, 2, \dots$), $\{h_{l\alpha}\}_{\alpha=1}^{\infty}$ and

1. See Appendix A for the derivation of the second equality in (4.6e). Note that in the second line of (4.6e) the assumption is made that N_i has full rank, in other words, that $m_i = r_i$. This assumption is essential in Sections 3.5 and 3.6 and in Appendix A, but can be dropped in the rest of Section 3.4.

$\{h_{u\alpha}\}_{\alpha=1}^{\infty}$ denote sequences in R^+, where $h_{l\alpha} < h_{u\alpha}$ ($\alpha = 1, 2, \ldots$), and $\{B_\alpha\}_{\alpha=1}^{\infty}$ denote a sequence of subsets ('blocks') of R^{r+1}, where

$$(5.1) \qquad B_\alpha = \{(\beta, h) | \beta \in R^r, h \in R^+, \beta_{l\alpha} < \beta < \beta_{u\alpha}, h_{l\alpha} < h < h_{u\alpha}\}$$

Furthermore, let $\{B_\alpha^*\}_{\alpha=1}^{\infty}$ denote a sequence of subsets of R^{r+1}, where

$$(5.2) \qquad B_\alpha^* = \{(\beta, \log h) | \beta \in R^r, h \in R^+, \beta_{l\alpha} < \beta < \beta_{u\alpha}, h_{l\alpha} < h < h_{u\alpha}\}$$

We assume that the components of β and $\log h$ are independently distributed, and specify the following prior densities for $\alpha = 1, 2, \ldots$

$$(5.3) \qquad f'_\alpha(\beta, \log h) = \gamma'_\alpha \qquad \text{if } (\beta, \log h) \in B_\alpha^*$$

$$= 0 \qquad \text{otherwise}$$

with

$$(5.4) \qquad 1/\gamma'_\alpha = \left\{\prod_{j=1}^{r} (\beta_{u\alpha j} - \beta_{l\alpha j})\right\} (\log h_{u\alpha} - \log h_{l\alpha})$$

where $\beta_{u\alpha j}$ denotes the j-th component of the r-dimensional vector $\beta_{u\alpha}$, and where $\beta_{l\alpha j}$ is similarly defined. Transformation of $\log h$ to h yields

$$(5.5) \qquad f'_\alpha(\beta, h) = \gamma'_\alpha h^{-1} \qquad \text{if } (\beta, h) \in B_\alpha$$

$$= 0 \qquad \text{otherwise}$$

Combining the prior densities (5.5) and the likelihood $l(y|\beta, h)$, we obtain the posterior densities which are products of a function depending only on α and a function independent of α. Hence we can write

$$(5.6) \qquad f''_\alpha(\beta, h) = \gamma''_\alpha \frac{|N|^{\frac{1}{2}} (\frac{1}{2}\lambda v)^{\frac{1}{2}\lambda} h^{\frac{1}{2}n-1}}{(2\pi)^{\frac{1}{2}r} (\frac{1}{2}\lambda - 1)!} \exp[-\tfrac{1}{2}h\{\lambda v + (\beta - b)^t N (\beta - b)\}]$$

$$\text{if } (\beta, h) \in B_\alpha$$

$$= 0 \qquad \text{otherwise}$$

where γ''_α can be calculated by means of

$$(5.7) \qquad \int_{B_\alpha} f''_\alpha(\beta, h) \, d\beta \, dh = 1$$

It is clear that γ''_α is approximately equal to unity if $\beta_{u\alpha}$, $-\beta_{l\alpha}$, $h_{u\alpha}$, and $1/h_{l\alpha}$ are taken large enough. More precisely, the sequence of posterior distributions obtained by considering sequences of values of $\beta_{u\alpha}$, $-\beta_{l\alpha}$, $h_{u\alpha}$, and $1/h_{l\alpha}$ that diverge to infinity converges in distribution to a Normal-gamma distribution with density

$$(5.8) \qquad f(\beta, h) = f_{N\gamma}^{(r)}(\beta, h | b, v, N, \lambda)$$

since the limit (5.8) of the sequence of posterior densities so obtained is a proper density;[1] see Scheffé (1947). It follows from (5.5) and (5.6) that we can write for the alternative models

$$(5.9) \qquad C_i' = \gamma_{\alpha i}'$$

$$(5.10) \qquad C_i'' = \frac{\gamma_{\alpha i}'' |N_i|^{\frac{1}{2}} (\frac{1}{2}\lambda_i v_i)^{\frac{1}{2}\lambda_i}}{(2\pi)^{\frac{1}{2}r_i} (\frac{1}{2}\lambda_i - 1)!}$$

compare (2.5) and (2.10). These results can be combined with (3.2e) and substituted in (2.9) to obtain

$$(5.11) \qquad \frac{p_{\alpha i}''}{p_{\alpha j}''} = \frac{p_i' \gamma_{\alpha i}' \gamma_{\alpha j}'' |N_j|^{\frac{1}{2}} (\frac{1}{2}\lambda_i - 1)! (\frac{1}{2}\lambda_j v_j)^{\frac{1}{2}\lambda_j} (2\pi)^{\frac{1}{2}r_i} |J_i|}{p_j' \gamma_{\alpha j}' \gamma_{\alpha i}'' |N_i|^{\frac{1}{2}} (\frac{1}{2}\lambda_j - 1)! (\frac{1}{2}\lambda_i v_i)^{\frac{1}{2}\lambda_i} (2\pi)^{\frac{1}{2}r_j} |J_j|}$$

If the limiting procedure described above is carried out, we find (for $i = 1, \ldots, k$) that $\gamma_{\alpha i}''$ tends to unity and $\gamma_{\alpha i}'$ tends to zero, which yields an indeterminate result for (5.11) unless additional assumptions are made about the ratio $\gamma_{\alpha i}'/\gamma_{\alpha j}'$. We consider it would be difficult, if not impossible, to defend a general assumption about this ratio, but in certain cases *ad hoc* assumptions might work well.

3.6. A LARGE SAMPLE PROPERTY

In this section, we leave the field of prior and posterior analysis and consider a property of the conditional sampling distribution of $(p_1'', \ldots, p_k'')$, given the true state θ. For simplicity, we assume

(i) X_i fixed for $i = 1, \ldots, k$; (ii) $|J_i| = |J_j|$ for $i = 1, \ldots, k$, and we confine ourselves to the probability limit of $(p_1'', \ldots, p_k'')$, as the number of observations tends to infinity. We shall prove that if $\bar{\theta}$ is the true state of nature and belongs to Θ_i, and if assumptions (i) and (ii) hold, we have

$$\operatorname*{plim}_{n \to \infty} p_i'' = 1 \qquad \operatorname*{plim}_{n \to \infty} p_j'' = 0 \qquad (j \neq i)$$

Without loss of generality, we assume throughout this section that model 1 is the right model and that some unknown pair $(\bar{\beta}_1, \bar{h}_1)$ represents the true

1. The existence of this integral follows from Fubini's Theorem; see, e.g., MUNROE (1959), Section 28.7.

state of nature, and we compare this model with some arbitrary model 2. We adopt the assumptions made in the first paragraph of Section 3.3, but confine ourselves to the case where X_1 and X_2 are fixed and the variables to be explained are equal for both models, so that $|J_2| = |J_1|$. Further, we need some assumptions about the behavior of the matrices of explanatory variables. Let us write X_{in} for the matrix X_i with n rows, and assume that X_{in} has full column rank (so that $(X_{in}^t X_{in})^{-1}$ exists) and that

$$(6.1) \qquad \lim_{n\to\infty} (X_{in}^t X_{in})^{-1} = 0 \qquad\qquad (i = 1, 2)$$

In addition, we assume that

$$(6.2) \qquad \lim_{n\to\infty} \frac{|X_{1n}^t X_{1n}|^{\frac{1}{2}}}{|X_{2n}^t X_{2n}|^{\frac{1}{2}}} \cdot \frac{1}{(1+\eta)^n} = 0$$

for some positive number η to be fixed below. Note that it is sufficient for (6.2) that real numbers $A > 0$ and μ exist such that

$$n^{\mu} \frac{|X_{1n}^t X_{1n}|^{\frac{1}{2}}}{|X_{2n}^t X_{2n}|^{\frac{1}{2}}} < A$$

for all values of n from some n_0 onward. One has to construct rather unusual X matrices to violate this condition.

Concerning the parameters of the natural-conjugate prior distribution, we remark that they are independent of n so that the probability limit of (4.7), if it exists, is equal to

$$(6.3) \qquad \operatorname*{plim}_{n\to\infty} \frac{p_2''}{p_1''} = C^* \operatorname*{plim}_{n\to\infty} \frac{|N_1''|^{\frac{1}{2}} (\frac{1}{2}\lambda_1'' v_1'')^{\frac{1}{2}\lambda_1''} (\frac{1}{2}\lambda_2'' - 1)!}{|N_2''|^{\frac{1}{2}} (\frac{1}{2}\lambda_2'' v_2'')^{\frac{1}{2}\lambda_2''} (\frac{1}{2}\lambda_1'' - 1)!}$$

where C^* is implicitly defined as

$$(6.4) \qquad C^* = \frac{p_2' |N_2'|^{\frac{1}{2}} (\frac{1}{2}\lambda_2' v_2')^{\frac{1}{2}\lambda_2'} (\frac{1}{2}\lambda_1' - 1)!}{p_1' |N_1'|^{\frac{1}{2}} (\frac{1}{2}\lambda_1' v_1')^{\frac{1}{2}\lambda_1} (\frac{1}{2}\lambda_2' - 1)!}$$

Note that C^* is independent of the sample size. First, let us consider the ratio $(\frac{1}{2}\lambda_2'' - 1)!/(\frac{1}{2}\lambda_1'' - 1)!$. If we write $\lambda_2'' = \lambda_1'' + 2\omega$, where ω is some real number independent of n, we find by Stirling's formula for sufficiently large values of λ_1'' and λ_2''

$$(6.5) \qquad \frac{(\frac{1}{2}\lambda_2'' - 1)!}{(\frac{1}{2}\lambda_1'' - 1)!} = \frac{(\frac{1}{2}\lambda_1'' + \omega - 1)!}{(\frac{1}{2}\lambda_1'' - 1)!} \approx \frac{\sqrt{\frac{1}{2}\lambda_1'' - 1 + \omega}\,\{(\frac{1}{2}\lambda_1'' - 1 + \omega)/\mathrm{e}\}^{\frac{1}{2}\lambda_1'' - 1 + \omega}}{\sqrt{\frac{1}{2}\lambda_1'' - 1}\,\;\{(\frac{1}{2}\lambda_1'' - 1)/\mathrm{e}\}^{\frac{1}{2}\lambda_1'' - 1}}$$

$$= \sqrt{1 + \frac{\omega}{\frac{1}{2}\lambda_1'' - 1}} \left\{1 + \frac{\omega}{\frac{1}{2}\lambda_1'' - 1}\right\}^{\frac{1}{2}\lambda_1'' - 1} \mathrm{e}^{-\omega} (\tfrac{1}{2}\lambda_1'' - 1 + \omega)^{\omega}$$

see, e.g., Courant (1937), Vol. 1, Chapter VII. If $n \to \infty$, also $\lambda_1'' \to \infty$ and $\lambda_2'' \to \infty$ (see (4.6d) and (3.3d)), so that (6.3) can be rewritten as

$$\text{(6.6)} \qquad \underset{n\to\infty}{\text{plim}} \frac{p_2''}{p_1''} = C^* \underset{n\to\infty}{\text{plim}} \frac{|N_1''|^{\frac{1}{2}}(\lambda_1'' v_1''/n)^{\frac{1}{2}\lambda_1''}}{|N_2''|^{\frac{1}{2}}(\lambda_2'' v_2''/n)^{\frac{1}{2}\lambda_2''}}$$

since

$$\text{(6.7)} \qquad \lim_{n\to\infty} \sqrt{1+\frac{\omega}{\frac{1}{2}\lambda_1''-1}} \left\{1+\frac{\omega}{\frac{1}{2}\lambda_1''-1}\right\}^{\frac{1}{2}\lambda_1''-1} e^{-\omega}(\tfrac{1}{2}\lambda_1''-1+\omega)^{\omega}(\tfrac{1}{2}n)^{-\omega} = 1$$

Next, let us consider

$$\text{(6.8)} \qquad \underset{n\to\infty}{\text{plim}} \frac{\lambda_1'' v_1''}{n} = \underset{n\to\infty}{\text{plim}} \frac{\lambda_1' v_1' + b_1''^t N_1' b_1' + \lambda_1 v_1 + b_1^t N_1 b_1 - b_1''^t N_1'' b_1''}{n}$$

$$= \underset{n\to\infty}{\text{plim}} \frac{\lambda_1 v_1}{n} + \underset{n\to\infty}{\text{plim}} \frac{(b_1-b_1')^t\{(N_1')^{-1}+N_1^{-1}\}^{-1}(b_1-b_1')}{n}$$

compare (4.6e). Recalling that $y_n = X_{1n}\bar{\beta}_1 + \varepsilon_n$, where ε_n is normally distributed with zero mean and covariance matrix $(1/\bar{h}_1)I_n$, and $b_1 = (X_{1n}^t X_{1n})^{-1} X_{1n}^t y_n$, it is well-known that

$$\text{(6.9)} \qquad \underset{n\to\infty}{\text{plim}} \frac{\lambda_1 v_1}{n} = \sigma^2$$

where $\sigma^2 = 1/\bar{h}_1$; see, e.g., Mood and Graybill (1963), p. 348. To find the probability limit of the second term of the final expression of (6.8), we start with b_1 and conclude that (6.1) is a necessary and sufficient condition for b_1 to converge in the squared mean to $\bar{\beta}_1$, since $Eb_1 = \bar{\beta}_1$ for each n, and

$$\lim_{n\to\infty} V(b_1) = \sigma^2 \lim_{n\to\infty} (X_{1n}^t X_{1n})^{-1} = 0$$

Convergence in the squared mean implies convergence in probability, so that

$$\text{(6.10)} \qquad \underset{n\to\infty}{\text{plim}}\, b_1 = \bar{\beta}_1$$

and

$$\text{(6.11)} \qquad \underset{n\to\infty}{\text{plim}} (b_1 - b_1') = \bar{\beta}_1 - b_1'$$

Next, given assumption (6.1), it is easy to verify that

$$\text{(6.12)} \qquad \lim_{n\to\infty} \{(N_1')^{-1} + N_1^{-1}\}^{-1} = N_1'$$

Combining (6.11) and (6.12), we obtain

$$(6.13) \quad \operatorname*{plim}_{n\to\infty} \frac{(b_1-b_1')^t\{(N_1')^{-1}+N_1^{-1}\}^{-1}(b_1-b_1')}{n} = 0$$

so that

$$(6.14) \quad \operatorname*{plim}_{n\to\infty} \frac{\lambda_1'' v_1''}{n} = \sigma^2$$

see (6.8) and (6.9).

Next, we continue with the second model and consider the limiting behavior of

$$(6.15) \quad \frac{\lambda_2'' v_2''}{n} = \frac{\lambda_2' v_2' + \lambda_2 v_2 + (b_2-b_2')^t\{(N_2')^{-1}+N_2^{-1}\}^{-1}(b_2-b_2')}{n}$$

Here $\lambda_2 v_2 = (y_n - X_{2n} b_2)^t (y_n - X_{2n} b_2)$ where $b_2 = (X_{2n}^t X_{2n})^{-1} X_{2n}^t y_n$, while, in fact, y has been generated by $y_n = X_{1n}\bar{\beta}_1 + \varepsilon_n$. We assume that there is *no* value of β_2 such that $X_{2n}\beta_2 = X_{1n}\bar{\beta}_1$ (otherwise the second specification would be a correct alternative to the first one). Then we can write

$$(6.16) \quad \frac{\lambda_2 v_2}{n} = \frac{\bar{\beta}_1^t X_{1n} M_{2n} X_{1n}\bar{\beta}_1 + 2\varepsilon_n^t M_{2n} X_{1n}\bar{\beta}_1 + \varepsilon_n^t M_{2n}\varepsilon_n}{n}$$

where M_{2n} is an idempotent and symmetric matrix defined by

$$(6.17) \quad M_{2n} = I - X_{2n}(X_{2n}^t X_{2n})^{-1} X_{2n}^t$$

Our assumption $X_{2n}\beta_2 \neq X_{1n}\bar{\beta}_1$ (all β_2) implies that $M_{2n}X_{1n}\bar{\beta}_1 \neq 0$. Now, if we assume[1] that, as the sample size increases, the sequence $(\bar{\beta}_1^t X_{1n}^t M_{2n} X_{1n}\bar{\beta}_1)/n$ is bounded below by a positive number a_1, it can be shown (see Kloek (1970)) that for each $\delta > 0$

$$\lim_{n\to\infty} P\left[\frac{\lambda_2 v_2}{n} \geqslant \sigma^2 + a_1 - \delta\right] = 1$$

If we choose an arbitrary value for δ such that $0 < \delta < a_1$ and define $a_2 = a_1 - \delta$, we obtain

$$(6.18) \quad \lim_{n\to\infty} P\left[\frac{\lambda_2 v_2}{n} \geqslant \sigma^2 + a_2\right] = 1$$

1. Note that $M_{2n}X_{1n}\bar{\beta}_1$ can be interpreted as the vector of least-squares residuals which results when $X_{1n}\bar{\beta}_1$ is 'explained' by the columns of X_{2n}. So the assumption says that the mean square of these residuals will never be smaller than a positive number a_1.

Concerning the quadratic form on the right hand-side of (6.15), we remark that we do not need an extensive analysis as was the case with model 1, since the matrix $\{(N_2')^{-1}+N_2^{-1}\}^{-1}$ is positive definite (also in the limiting case) and thus the quadratic form is positive for all n. This implies that

$$(6.19) \qquad \lim_{n\to\infty} P\left[\frac{\lambda_2'' v_2''}{n} \geqslant \sigma^2 + a_2\right] = 1$$

see (6.15) and (6.18).

We go on to investigate the ratio of determinants of (6.6) and consider

$$(6.20) \qquad \frac{|N_1''|^{\frac{1}{2}}}{|N_2''|^{\frac{1}{2}}} = \frac{|N_1'+N_1|^{\frac{1}{2}}}{|N_2'+N_2|^{\frac{1}{2}}}$$

$$= \frac{|X_{1n}^t X_{1n}|^{\frac{1}{2}} |(X_{1n}^t X_{1n})^{-1} N_1' + I|^{\frac{1}{2}}}{|X_{2n}^t X_{2n}|^{\frac{1}{2}} |(X_{2n}^t X_{2n})^{-1} N_2' + I|^{\frac{1}{2}}}$$

Note that, if $n \to \infty$, the right-hand determinants in the numerator as well as in the denominator tend to 1 since the corresponding matrices converge to unity matrices; compare (6.1). Collecting (6.14), (6.19), and (6.20), we obtain for (6.6)

$$(6.21) \qquad \operatorname*{plim}_{n\to\infty} \frac{p_2''}{p_1''} \leqq C^* \lim_{n\to\infty} \frac{|X_{1n}^t X_{1n}|^{\frac{1}{2}} (\sigma^2)^{\frac{1}{2}\lambda_1''}}{|X_{2n}^t X_{2n}|^{\frac{1}{2}} (\sigma^2 + a_2)^{\frac{1}{2}\lambda_2''}}$$

$$= C^* \sigma^{-2\omega} \lim_{n\to\infty} \frac{|X_{1n}^t X_{1n}|^{\frac{1}{2}}}{|X_{2n}^t X_{2n}|^{\frac{1}{2}}} \cdot \frac{1}{(1 + a_2/\sigma^2)^{\frac{1}{2}\lambda_2''}}$$

$$= 0$$

if (6.2) is assumed to hold true for some $\eta \geqslant a_2/\sigma^2$. This implies

$$(6.22) \qquad \operatorname*{plim}_{n\to\infty} p_1'' = 1 \text{ and } \operatorname*{plim}_{n\to\infty} p_2'' = 0; \qquad \text{q.e.d.}$$

By investigating the limiting behavior of (5.11) and accepting the assumption that the ratio $\gamma_{\alpha 2}'/\gamma_{\alpha 1}'$ is bounded, we can verify along the same lines that the previous assumptions and results can be used to find that (6.22) also holds if we start from (5.11).

3.7. PRELIMINARY REMARKS ABOUT APPLICABILITY

So far we have derived a number of mathematical results. It is too early to discuss fully whether and to what extent these results are applicable. We shall therefore confine ourselves to six remarks.

(i) Of course, the large-sample results in Section 3.6 imply that for any values of the prior parameters our discriminating method will select with near-certainty the correct model – provided there is one in the set of models considered – as long as our sample size is sufficiently large. It should be noted that we have confined ourselves in Section 3.6 to the comparison of models with the same variable to be explained. Ratios of Jacobians unequal to unity are much more difficult to handle.

(ii) If the investigator is able and willing to specify proper natural-conjugate prior densities on all the models he considers, he can (in principle) apply (4.7) without any difficulty. He should be warned, however, that this specification problem is not at all simple. Suppose, for example, that he wants to compare two models which he considers equally probable *a priori*. He may express this opinion by setting $p_1' = p_2' = \frac{1}{2}$. But then his final result p_1''/p_2'' may be quite sensitive to the way he specifies b_i', N_i', λ_i' and v_i' $(i = 1, 2)$; compare (4.6) and (4.7). So we advise anyone who uses (4.7) to carry out some sensitivity analysis for acceptable variations of the prior parameters. Some of the most obvious possibilities for practical application are discussed in remark (iv) below.

(iii) In case the investigator has no information (or very little) on β and h, he may wish to consider the limiting behavior of (5.11) as α tends to infinity. As it has already been mentioned, it will turn out that no unique limit of $\gamma_{\alpha i}'/\gamma_{\alpha j}'$ can be obtained unless a specific assumption is made about 'how fast' these quantities tend to zero in comparison with each other. It is impossible to give general rules here, but there are some particular cases in which rather natural solutions are available. For these we refer to the following remark.

(iv) We now briefly describe some sets of alternative models for which the comparability problems for both natural-conjugate and uniform prior distributions are relatively small. The larger part of these problems will be discussed in more detail in the remainder of this book. Let, for example, the models differ only as to alternative assumptions about the disturbances, such as homoskedasticity versus heteroskedasticity, or independent versus Markov-autocorrelated disturbances. In such cases, it seems acceptable to specify identical prior distributions on β, and no serious problems arise in specifying the prior parameters pertaining to the precision; see Chapter 4. Other examples for which we treat the problems of specifying the parameters in a consistent way are the comparison of a level model with its first-differences model, and that of a set of distributed-lag models, where the most probable lag parameter must be found; see Chapter 4 and Chapter 5, respectively. As a third example, although we do not go into this subject in this book, we mention the comparison of alternative proxy-variables for explanatory variables which can not be measured. Also in such a case one may often assume that the prior parameters are independent of the way in which the explanatory variables are approximated. For examples of such a problem, we refer to a paper by Harkema and Kloek (1969), where this subject is treated by means of the natural-conjugate approach, and to Geisel (1970) who deals with a non-informative approach as well as with Beta distributed parameters. The paper of Harkema and Kloek also illustrates that, for purposes of prediction and decision, one is not forced to choose a single model, but may proceed with a mixture of all or some conditional posterior distributions.

(v) For the sake of completeness, we note that the comparison of models with the same dependent variable and different explanatory variables, is, as shall be pointed out, not simple at all. This same conclusion can be drawn with respect to choosing among models with different but functionally related dependent variables. Chapters 5 and 6 mainly deal with these two less straightforward applications of posterior probabilities.

(vi) Finally, in the case that application of a non-informative uniform prior distribution to a single model yields a posterior mean for one of the parameters whose sign or magnitude is not in accordance with the investigator's prior beliefs, some prior knowledge can be incorporated without using the natural-conjugate approach. This possibility is discussed in the next section.

3.8. TRIAL AND ERROR IN PRIOR SPECIFICATION

As has already been discussed in Section 2.3 many econometricians are reluctant to specify definite prior ideas about the parameters of their model. In many cases, however, they have some prior knowledge about signs or magnitudes of these parameters. Then, the following approach can be used. Let us consider the linear model

$$z = X\beta + \varepsilon \tag{8.1}$$

and accept the non-informative uniform prior distribution as discussed in Section 2.6. Then the marginal posterior distribution of β is Student with mean and variance

$$\begin{aligned} E(\beta) &= (X^t X)^{-1} X^t z = b \\ V(\beta) &= (\lambda v/(\lambda - 2))(X^t X)^{-1} \end{aligned} \tag{8.2}$$

Suppose that one of the elements of b, say $b(1)$, has a wrong sign. This implies that the resulting posterior distribution disagrees with the econometrician's best judgment. Raiffa and Schlaifer (1961), Section 3.3.3, state that for such a case it is irresponsible to hold to the original prior distribution. A possible solution to this problem is to specify some acceptable values of $\beta(1)$, e.g., $\beta_1'(1)$, $\beta_2'(1)$, ..., $\beta_k'(1)$ with prior probabilities $p_1', p_2', \ldots, p_k'$. Then it is possible to consider k different models

$$(8.3) \qquad y_i \equiv z - \beta_i(1) x_1 = X^* \beta^* + \varepsilon \qquad i = 1, \ldots, k$$

where X^* and β^* are obtained by deleting x_1 and $\beta(1)$, respectively. The application of (5.11) to (8.3) in order to obtain $p_1'', p_2'', \ldots, p_k''$ is not straightforward, since some additional assumptions still have to be made. We accept non-informative prior distributions on β^* and $\log h$. Since β^* and h occur in every model, and since we assume that all elements of β and h are independently distributed, we can accept equal uniform prior distributions for β^* and h so that $\gamma_{\alpha i}'/\gamma_{\alpha j}'$ equals one for all α and $i, j = 1, \ldots, k$. By assumption (2.4.6) it can be verified that $|J_i| = |J_j|$ for all i, j. Given these assumptions it follows that (5.11) can be written as

$$(8.4) \qquad \frac{p_i''}{p_j''} = \lim_{\alpha \to \infty} \frac{p_{i\alpha}''}{p_{j\alpha}''} = \frac{p_i' v_j^{\frac{1}{2}\lambda}}{p_j' v_i^{\frac{1}{2}\lambda}}$$

where

$$\lambda = n - r + 1, \; v_i = (y_i - X^* b_i^*)^t (y_i - X^* b_i^*)/\lambda,$$

and

$$b_i^* = (X^{*t} X^*)^{-1} X^{*t} y_i.$$

The posterior probabilities of (8.4) define the discrete marginal posterior distribution of $\beta(1)$. The mean and variance are

$$(8.5) \qquad \begin{aligned} E(\beta(1)) &= \sum_{i=1}^{k} p_i'' \beta_i'(1) \\ V(\beta(1)) &= \sum_{i=1}^{k} p_i'' (\beta_i'(1))^2 - (E(\beta(1)))^2 \end{aligned}$$

For the elements β^* one may proceed with a weighted mixture of the marginal posterior distributions of β^*, conditional upon $\beta_i'(1)$. In other words, the marginal posterior distribution of β^* is a weighted sum of Student distributions with the following moments

$$(8.6) \qquad \begin{aligned} E(\beta^*) &= \sum_{i=1}^{k} p_i'' b_i^* \\ V(\beta^*) &= \sum_{i=1}^{k} p_i'' \{(\lambda v_i/(\lambda - 2))(X^{*t} X^*)^{-1} + E(\beta_i^*) E(\beta_i^*)^t\} + \\ &\quad - E(\beta^*) E(\beta^*)^t \end{aligned}$$

It is clear that if the means of the k different Student distributions differ by large amounts, the moments of (8.6) are difficult to interpret. If, however, one of the posterior probabilities of (8.4) is large in comparison with the others, one may decide to continue with this single model.

It is needless to say that if the new posterior means of the elements of β are not in accordance with our prior knowledge, this procedure can be repeated until all unacceptable posterior means have disappeared. In other words, generalization of the method proposed is straightforward.

APPENDIX A

In this appendix we derive the second equality of (3.4.6e). We neglect the subscripts, since the result holds for any of the models considered provided the N matrix has full rank. We make repeated use of the following property, which holds for arbitrary non-singular matrices A and B of the same order:

$$\text{(A.1)} \qquad (A+B)^{-1} = B^{-1}(B^{-1}+A^{-1})^{-1}A^{-1} = A^{-1}(A^{-1}+B^{-1})^{-1}B^{-1}$$

Our starting point is $\lambda''v''$ and we find from (3.4.6e) that

$$\text{(A.2)} \qquad \lambda''v'' = \lambda'v' + b'^t N' b' + \lambda v + b^t Nb - b''^t N'' b''$$

We proceed by rewriting the terms of (A.2) containing b, b', and b'', as follows

$$\begin{aligned}
\text{(A.3)} \qquad & b'^t N' b' + b^t Nb - b''^t N'' b'' \\
&= b'^t N' b' + b^t Nb - (N'b' + Nb)^t (N'+N)^{-1}(N'b'+Nb) \\
&= b'^t N' b' + b^t Nb - b'^t N'(N'+N)^{-1}N'b' - b^t N(N'+N)^{-1}Nb + \\
&\quad - b'^t N'(N'+N)^{-1}Nb - b^t N(N'+N)^{-1}N'b' \\
&= b'^t N'\{I-(N'+N)^{-1}N'\}b' + b^t N\{I-(N'+N)^{-1}N\}b + \\
&\quad - 2b^t N(N'+N)^{-1}N'b' \\
&= b'^t N'(N'+N)^{-1}Nb' + b^t N(N'+N)^{-1}N'b + \\
&\quad - 2b^t N(N'+N)^{-1}N'b' \\
&= b'^t\{N^{-1}+(N')^{-1}\}^{-1}b' + b^t\{N^{-1}+(N')^{-1}\}^{-1}b + \\
&\quad - 2b^t\{N^{-1}+(N')^{-1}\}^{-1}b' \\
&= (b-b')^t\{(N')^{-1}+N^{-1}\}^{-1}(b-b')
\end{aligned}$$

Note that use has been made of equalities of the following type

$$I-(N'+N)^{-1}N' = (N'+N)^{-1}(N'+N)-(N'+N)^{-1}N' = (N'+N)^{-1}N$$

Now $\lambda''v''$ (compare (A.2)) can be written as

$$\text{(A.4)} \qquad \lambda''v'' = \lambda'v' + \lambda v + (b-b')^t\{(N')^{-1}+N^{-1}\}^{-1}(b-b')$$

4. Nonspherical disturbances

4.1. INTRODUCTION

In this chapter we apply the Bayesian procedure of posterior probabilities of alternative linear models, as discussed in Chapter 3, for the purpose of comparing linear models with alternative nonspherical assumptions concerning the disturbances.[1] In one of the simplest versions of the linear model $y = X\beta + \varepsilon$ one assumes that the disturbances are independent and normally distributed with $E\varepsilon = 0$ and $E\varepsilon\varepsilon^t = \sigma^2 I$. The most obvious generalization of these assumptions is that $E\varepsilon\varepsilon^t = \sigma^2 \Omega$, where Ω is a real positive definite symmetric (PDS) matrix. This Ω matrix, however, is not manageable without restrictions, since it contains too many parameters. Therefore, one usually considers simple cases as diagonal or first-order Markov-scheme Ω matrices.

The simple assumption $E\varepsilon\varepsilon^t = \sigma^2 I$ implies that all disturbances have the same variance σ^2. This property is often denoted by the term *homoskedastic* disturbances. For many purposes this assumption is too restrictive. High-income families, for example, show much greater variability in their saving behavior than low-income families, so that the assumption of common disturbance variance would be inappropriate for a cross-section saving-income relationship. One can deal with this phenomenon by assuming that the standard deviation of the i-th disturbance is proportional to the income (per head or equivalent adult) of the i-th family. More generally, in cross-section analysis one can assume that the standard deviation of the i-th disturbance is proportional to the i-th value of some explanatory variable, or with the i-th element of $X\alpha$ for some given vector α. Similar assumptions may be made in time-series analysis, or (alternatively) in a growing economy one may assume that the standard deviation of the j-th disturbance is proportional to $(1+g)^j$, where g is an average growth rate. If the disturbances have different variances, they are said to be *heteroskedastic*. Now we adopt the

1. Parts of this chapter are based on a paper by LEMPERS and KLOEK (1970).

assumption of heteroskedasticity but retain the independence assumption, so that our basic linear model can be rewritten as

(1.1) $\quad y = X\beta + \varepsilon$

with

(1.2) $\quad E\varepsilon = 0 \quad \text{and} \quad E\varepsilon\varepsilon^t = \sigma^2 \Omega$

where Ω is a diagonal matrix with positive diagonal elements according to a pattern that is chosen in advance. Note that if we wish to compare a number of different possibilities, say k, k different matrices Ω_i $(i = 1, \ldots, k)$ are supposed to be specified. One of these matrices could, of course, be the unit matrix. This enables us to compare different assumptions concerning the heteroskedasticity with each other, and in addition, with the assumption of homoskedasticity.

Another simple alternative to the assumption $E\varepsilon\varepsilon^t = \sigma^2 I$ is the first-order Markov-scheme model which generates autocorrelated disturbances in a simple way. This subject has received much attention from econometricians. The Durbin-Watson (1950) article has become a classic, and Theil (1965) and Koerts (1967) developed the BLUS test. More recently, Abrahamse and Koerts (1969) derived a 'new class of disturbance estimators', which were applied to the problem of testing for auto-correlation by Abrahamse and Louter (1969). All these procedures consider two hypotheses, namely, zero autocorrelation against positive (or negative) autocorrelation. If the hypothesis of zero autocorrelation has to be rejected, then the amount of autocorrelation in the model in question is not yet known. The parameter of the autoregressive structure can be estimated by means of the maximum-likelihood method or, much simpler, by means of a two-stage estimation procedure as has been suggested by Durbin (1960).

Bayesian statisticians can handle this problem by assuming a uniform prior distribution in regard to the autocorrelation coefficient (denoted by ρ) and derive a posterior distribution of ρ, given the sample. See, for instance, Zellner and Tiao (1964) and Thornber (1967). In this chapter we construct k different models by choosing alternative values for the autocorrelation coefficient ρ. We do this by taking a finite number (k) of ρ values from the open interval $(-1, 1)$. It goes without saying that no continuous posterior distribution of ρ is obtained in this way, but that a continuous distribution may be approximated to any desired degree of accuracy, by taking a sufficiently large value for k.

The order of discussion is as follows. In Section 4.2 we start to discuss the linear transformations which send nonspherical disturbance vectors into spherical ones and the resulting transformations of the sample statistics.

We continue to simplify the formula of the posterior probabilities for the case of uniform prior distributions presented in Section 3.5, and also give a general approach to the problem of obtaining some definite value of the ratio of two normalizing constants of such uniform prior distributions for the cases we consider. We finish this section by deriving the ratio of posterior probabilities of two models with alternative assumptions concerning the pattern of the heteroskedastic disturbances. In Section 4.3 we consider the properties of the autocorrelated disturbances and derive their posterior probabilities for the case of uniform prior distributions and illustrate these results with some numerical examples in Section 4.4. In Section 4.5 we present some concluding remarks about the results we obtained. Section 4.6 deals with the problem of treating nonspherical disturbances in the case of natural-conjugate prior distributions, and some sensitivity analysis has been carried out in Section 4.7. Finally, there is Appendix B which contains the application of posterior probabilities to first-differences models.

4.2. THE CASE OF UNIFORM PRIOR DISTRIBUTIONS

To avoid notational confusion, and to obtain notational correspondence to Chapter 3, we rewrite the linear models we consider as

(2.1) $\qquad z = W\beta + \eta_i \qquad (i = 1, \ldots, k)$

where η_i is a normally distributed n-dimensional vector with

(2.2) $\qquad E\eta_i = 0$ and $E\eta_i\eta_i^t = h_i^{-1}\Omega_i$

It can easily be seen that the alternative models differ only in the assumptions made concerning the disturbances. This implies $r_i = r$ and $\lambda_i = \lambda$ $(i = 1, \ldots, k)$. If we transform such a model by means of a matrix B_i satisfying $B_i^t B_i = \Omega_i^{-1}$, we obtain

(2.3) $\qquad B_i z = B_i W\beta + B_i\eta_i$

or

(2.4) $\qquad y_i = X_i\beta + \varepsilon_i$

where $y_i = B_i z$, $X_i = B_i W$, $\varepsilon_i = B_i\eta_i$, and conclude that ε_i is normally distributed with $E\varepsilon_i = 0$ and $E\varepsilon_i\varepsilon_i^t = h_i^{-1}I$. Now the likelihood function as discussed in Section 3.3, is characterized by

(2.5a) $\qquad N_i = X_i^t X_i = W^t B_i^t B_i W = W^t\Omega_i^{-1}W$

(2.5b) $\qquad b_i = (X_i^t X_i)^{-1}X_i^t y_i = (W^t\Omega_i^{-1}W)^{-1}W^t\Omega_i^{-1}z$

if the matrix W has full column rank[1] r, and

(2.5c) $\lambda = n-r$

(2.5d) $\lambda v_i = (y_i - X_i b_i)^t (y_i - X_i b_i) = (z - Wb_i)^t \Omega_i^{-1} (z - Wb_i)$

(2.5e) $|J_i| = |B_i| = |\Omega_i|^{-\frac{1}{2}}$

where n is the number of observations.

We recall that our purpose is to compare alternative Ω matrices to obtain the most probable one in a given class with the help of the Bayesian procedure of posterior probabilities of alternative linear models. We shall make use of the results of Section 3.5. Here we confine ourselves to the case of uniform prior distributions, where β and $\log h$ $(h = \sigma^{-2})$ are uniformly and independently distributed. For such prior distributions the ratio of posterior probabilities of two alternative specifications, say i and j, is given by (3.5.11). In the heteroskedastic case, as well as in the first-order Markov-scheme case, we consider a set of models with different Ω matrices, all other assumptions being equal. As already mentioned we have in both cases $r_i = r_j = r$ and $\lambda_i = \lambda_j = \lambda$ $(i, j = 1, \ldots, k)$, so that the expression (3.5.11) can be considerably simplified, as follows

$$(2.6) \qquad \frac{p''_{\alpha i}}{p''_{\alpha j}} = \frac{\gamma'_{\alpha i}\gamma''_{\alpha j} |W^t \Omega_j^{-1} W|^{\frac{1}{2}} v_j^{\frac{1}{2}(n-r)} |\Omega_j|^{\frac{1}{2}}}{\gamma'_{\alpha j}\gamma''_{\alpha i} |W^t \Omega_i^{-1} W|^{\frac{1}{2}} v_i^{\frac{1}{2}(n-r)} |\Omega_i|^{\frac{1}{2}}}$$

where use has been made of (2.5a)–(2.5e) and where it has been assumed that $p'_i = p'_j$ to express the opinion that we consider all models equally probable *a priori.*

The problem of assigning uniform prior distributions to β and $\log h$, such that we obtain a determinate value for the ratio $\gamma'_{\alpha i}/\gamma'_{\alpha j}$, has a rather natural solution for the models we consider. We observe that there is no difference in the explanatory variables and variable to be explained; compare (2.1). Therefore, since the uniform prior distribution of the parameter β is taken independent of the prior distribution of the precision (h) of a model, we immediately conclude that it is natural to take the uniform prior distributions of the parameter β identical for the models we consider. We now turn to the distributions of the precisions of the models. Our conclusions will be equally simple but they are reached in a less straightforward way. Without loss of generality we restrict ourselves to two models i and j with precisions h_i and h_j. Then the problem arises as to how prior distributions of h_i and h_j should be specified. A particularly interesting

1. Note that this assumption is essential in view of (2.6), but that it can be dropped in Section 4.6. If rank $[W] < r$, then b_i denotes any solution of $(W^t \Omega_i^{-1} W) b_i = W^t \Omega_i^{-1} z$.

property is the following: the variance matrix $h^{-1}\Omega$ does not change when one replaces Ω by $\phi\Omega$ and at the same time h by ϕh, ϕ being a positive scalar. It is easily verified that the constant $1/(\log h_{u\alpha} - \log h_{l\alpha})$ (compare (3.5.4)) of the marginal prior density of $\log h$ is not changed by this transformation and it will be proved later that the posterior probabilities are not changed either. Therefore, we do not see much reason to bother about the ratio $\gamma'_{\alpha i}/\gamma'_{\alpha j}$ and simply assume that $\gamma'_{\alpha i}/\gamma'_{\alpha j} = 1$ for all α, i and j. In Section 3.5 we derived that $\lim_{\alpha\to\infty} \gamma''_{\alpha i} = 1$ $(i = 1, \ldots, k)$, so that the ratio of the posterior probabilities for $\alpha \to \infty$ can be written as

$$(2.7) \qquad \frac{p''_i}{p''_j} = \lim_{\alpha\to\infty} \frac{p''_{\alpha i}}{p''_{\alpha j}} = \frac{|W^t \Omega_j^{-1} W|^{\frac{1}{2}} v_j^{\frac{1}{2}(n-r)} |\Omega_j|^{\frac{1}{2}}}{|W^t \Omega_i^{-1} W|^{\frac{1}{2}} v_i^{\frac{1}{2}(n-r)} |\Omega_i|^{\frac{1}{2}}}$$

which can easily be computed. Note that the posterior probabilities of (2.7) are invariant against multiplication of one of the Ω matrices, say Ω_i, by an arbitrary positive number; compare (2.5). Finally we note that the application of (2.7) to the case of two heteroskedastic models is straightforward. Its application to the case of two first-order Markov-scheme models will be discussed in the next section.

4.3. FIRST-ORDER MARKOV-SCHEME MODELS

In this section we consider the properties of the disturbances of the linear model $z = W\beta + \eta$, where η is an n-dimensional vector of disturbances generated by a first-order Markov-scheme

$$(3.1) \qquad \eta(t) = \rho\eta(t-1) + \zeta(t) \qquad (t = 2, \ldots, n)$$

where[1] $|\rho| < 1$, and the $\zeta(t)$ are independent normally distributed random variables with zero mean and variance σ_ζ^2. Hence

$$(3.2) \qquad E\zeta = 0 \text{ and } E(\zeta\zeta^t) = \sigma_\zeta^2 I$$

where ζ is a column vector defined by

$$(3.3) \qquad \zeta^t = [\zeta(2) \ldots \zeta(n)]$$

For simplicity we add some assumptions on $\eta(1)$: $\eta(1), \zeta(2), \ldots, \zeta(n)$ have a joint normal distribution, with moments given by (3.2) and[2]

$$(3.4) \qquad E\eta(1) = 0$$

1. Note that ZELLNER and TIAO (1964) consider explosive ($|\rho| \geq 1$) as well as non-explosive ($|\rho| < 1$) schemes.
2. Note that (3.6) is tantamount to assuming that $\eta(1), \ldots, \eta(n)$ are homoskedastic.

(3.5) $E\eta(1)\zeta = 0$

(3.6) $\text{var}\,\eta(1) = \sigma_\zeta^2/(1-\rho^2) \equiv \sigma^2$

It follows that

(3.7) $\eta(t) = \rho^{t-1}\eta(1) + \sum_{s=0}^{t-2} \rho^s \zeta(t-s)$

(3.8) $E\eta(t) = 0$

(3.9) $\text{var}\,\eta(t) = \sigma^2$

(3.10) $E\eta(t)\eta(s) = \rho^{|t-s|}\sigma^2$

The complete variance matrix is as follows

$$(3.11) \quad E(\eta\eta^t) = \sigma^2 \begin{bmatrix} 1 & \rho & \rho^2 & \dots & \rho^{n-1} \\ \rho & 1 & \rho & \dots & \rho^{n-2} \\ \rho^2 & \rho & 1 & \dots & \rho^{n-3} \\ \dots & \dots & \dots & \dots & \dots \\ \rho^{n-1} & \rho^{n-2} & \rho^{n-3} & \dots & 1 \end{bmatrix} = \sigma^2\,\Omega \equiv h^{-1}\,\Omega$$

The determinant of Ω is given by $|\Omega| = (1-\rho^2)^{n-1}$ and its inverse is equal to

$$(3.12) \quad \Omega^{-1} = (1-\rho^2)^{-1} \begin{bmatrix} 1 & -\rho & 0 & \dots & 0 & 0 \\ -\rho & 1+\rho^2 & -\rho & \dots & 0 & 0 \\ 0 & -\rho & 1+\rho^2 & \dots & 0 & 0 \\ \dots & \dots & \dots & \dots & \dots & \dots \\ 0 & 0 & 0 & \dots & 1+\rho^2 & -\rho \\ 0 & 0 & 0 & \dots & -\rho & 1 \end{bmatrix}$$

Now if we transform the model $z = W\beta + \eta$ by means of a matrix B satisfying $B^t B = \Omega^{-1}$, we obtain

(3.13) $y = X\beta + \varepsilon$

where $y = Bz$, $X = BW$ and $\varepsilon = B\eta$, and conclude that ε is normally distributed with $E\varepsilon = 0$ and $E\varepsilon\varepsilon^t = h^{-1}I$. Given the results of Section 2 and denoting Ω_i as the Ω matrix for $\rho = \rho_i$, we find that the ratio of posterior probabilities for $\alpha \to \infty$ equals

$$(3.14) \quad \frac{p_i''}{p_j''} = \lim_{\alpha\to\infty} \frac{p_{\alpha i}''}{p_{\alpha j}''} = \frac{|W^t\Omega_j^{-1}W|^{\frac{1}{2}} v_j^{\frac{1}{2}(n-r)} (1-\rho_j^2)^{\frac{1}{2}(n-1)}}{|W^t\Omega_i^{-1}W|^{\frac{1}{2}} v_i^{\frac{1}{2}(n-r)} (1-\rho_i^2)^{\frac{1}{2}(n-1)}}$$

for $i, j = 1, \dots, k$.

4.4. NUMERICAL ILLUSTRATIONS

To illustrate the results of the previous section we generated a sample according to the following model

$$(4.1) \qquad z = 2.5 + 0.6w_1 - 0.94w_2 + \eta$$

where w_1 refers to the logarithm of an index number of real income per head and w_2 to the logarithm of a deflated price index number of spirits. Both vectors w_1 and w_2 were taken from the well-known spirits example due to Prest (1949) that deals with the demand for spirits in the United Kingdom from 1870 to 1938. The numbers 0.6 and -0.94 have been arbitrarily chosen; they represent the income elasticity and price elasticity, respectively. The disturbances were generated according to the model of the previous section with $\sigma = 0.15$ and $\rho = 0.6$. This standard deviation is rather large in comparison with the systematic part of (4.1) which is about 1.8, since all explanatory variables are about 2. It is also large in comparison with the estimated standard deviation 0.06 resulting from the application of least squares to the original sample.

We computed the posterior probabilities for 49 alternative values of ρ. These values were determined with the following scheme

$$(4.2) \qquad \rho_j = -1 + 0.04j \qquad j = 1, \ldots, 49$$

We also computed these posterior probabilities for different numbers of sample observations, namely 10, 21, 45, and 69. In Figure 1 below a graphical representation of the results can be found. The posterior probabilities for values of ρ smaller than zero are deleted since they were negligible in comparison with the others. Note that for the smallest sample ($n = 10$) no serious conclusion about the value of ρ can be drawn. A limited simulation experiment, however, suggests that, if we choose in advance a smaller standard deviation to generate the disturbances, the resulting pictures show larger mass concentrations around the true value of ρ. We recall that our posterior distributions of ρ are discrete. In order to make the figure clear, however, we have connected the points of the mass functions by continuous curves.

Further, in Table 4.1, one can find the modes as well as the (approximated) means of the posterior distributions of ρ. This table also contains Durbin-Watson statistics based on the residuals obtained by applying least squares to the simple model (2.1) with $\rho = 0$ and the transformed model (2.4) with the modal value of ρ, respectively. The fact that the most probable values of ρ are positive is in accordance with the values found for the Durbin-Watson statistics for $\rho = 0$, which are (except for the smallest sample)

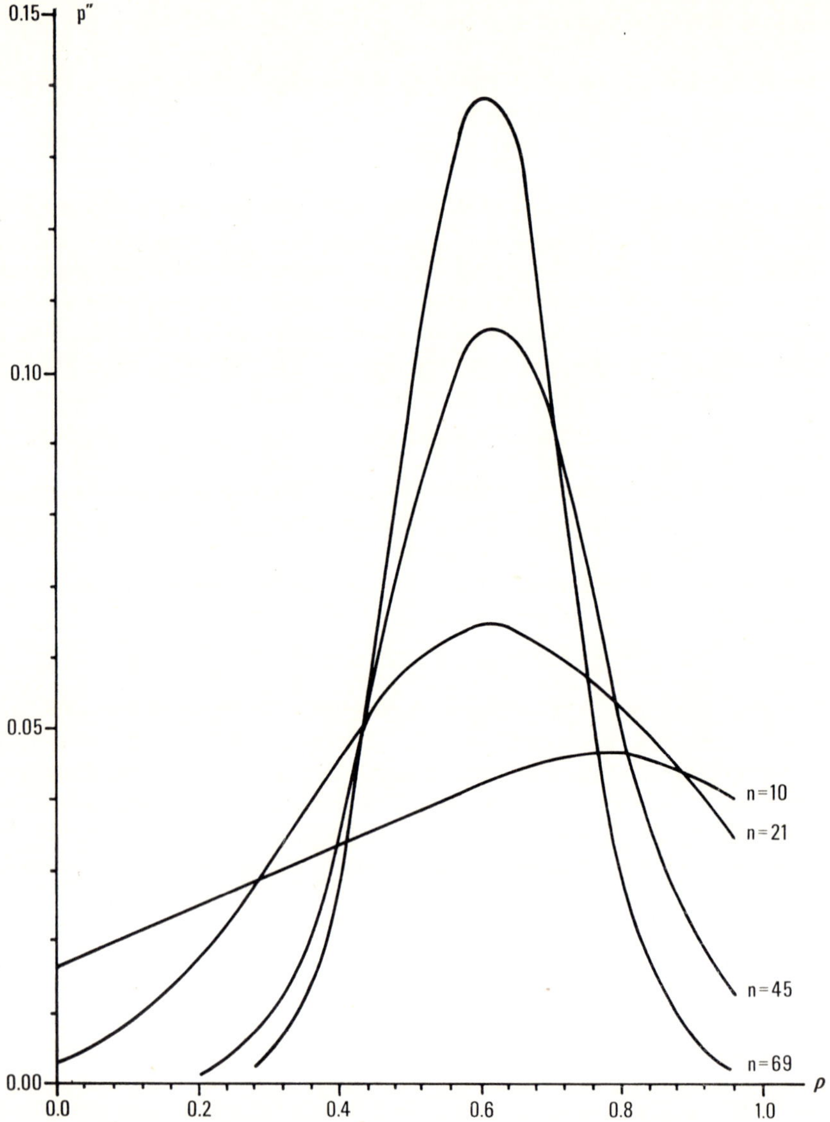

Fig. 1. *Posterior probabilities for positive values of* ρ *and different numbers of sample observations for the spirits example with artificial data*

considerably less than 2. The Durbin-Watson statistics for the modes of the posterior distributions of ρ are close to 2, which suggests that the autocorrelation has been removed; see Table 4.1.

TABLE 4.1. MODES AND MEANS OF POSTERIOR DISTRIBUTIONS OF ρ AND DURBIN-WATSON STATISTICS FOR FOUR SAMPLE SIZES

	Posterior Distribution of ρ		*Durbin-Watson Statistic for*	
n	*Mode*	*Mean*	$\rho = 0$	*Mode of ρ*
10	0.80	0.46	1.8039	2.1586
21	0.64	0.58	1.2143	2.1852
45	0.60	0.62	0.9709	2.1432
69	0.60	0.60	0.9338	1.9150

Finally, we also applied the procedure of posterior probabilities to the sample of the spirits example which was actually[1] observed. The Durbin-Watson statistic for $\rho = 0$ is equal to 0.2488 which indicates a high positive autocorrelation. After refinement of the alternative values of ρ up to three decimal places, the greatest posterior probability is obtained for $\rho = 0.989$, with (approximated) mean 0.979; see Figure 2. For these values the Durbin-Watson statistics are 2.0619 and 2.0344, respectively. Using $\rho = 0.989$ instead of $\rho = 0$, the posterior means of the income elasticity and price elasticity are 0.609 and -0.932, respectively; these results are more plausible than -0.120 and -1.228 for $\rho = 0$. It is worthwhile to note that Theil and Nagar (1961) obtained an estimate of ρ equal to 0.879. The posterior probability for this value, however, is negligible in comparison with the posterior probability for $\rho = 0.989$.

4.5. A COMPARISON BETWEEN DISCRETE AND CONTINUOUS PRIOR DISTRIBUTIONS

The numerical results of the previous section suggest that the procedure of posterior probabilities gives satisfactory results. For comparison of these

1. Then z refers to the index number of the logarithm of quantity (of spirits) demanded per head.

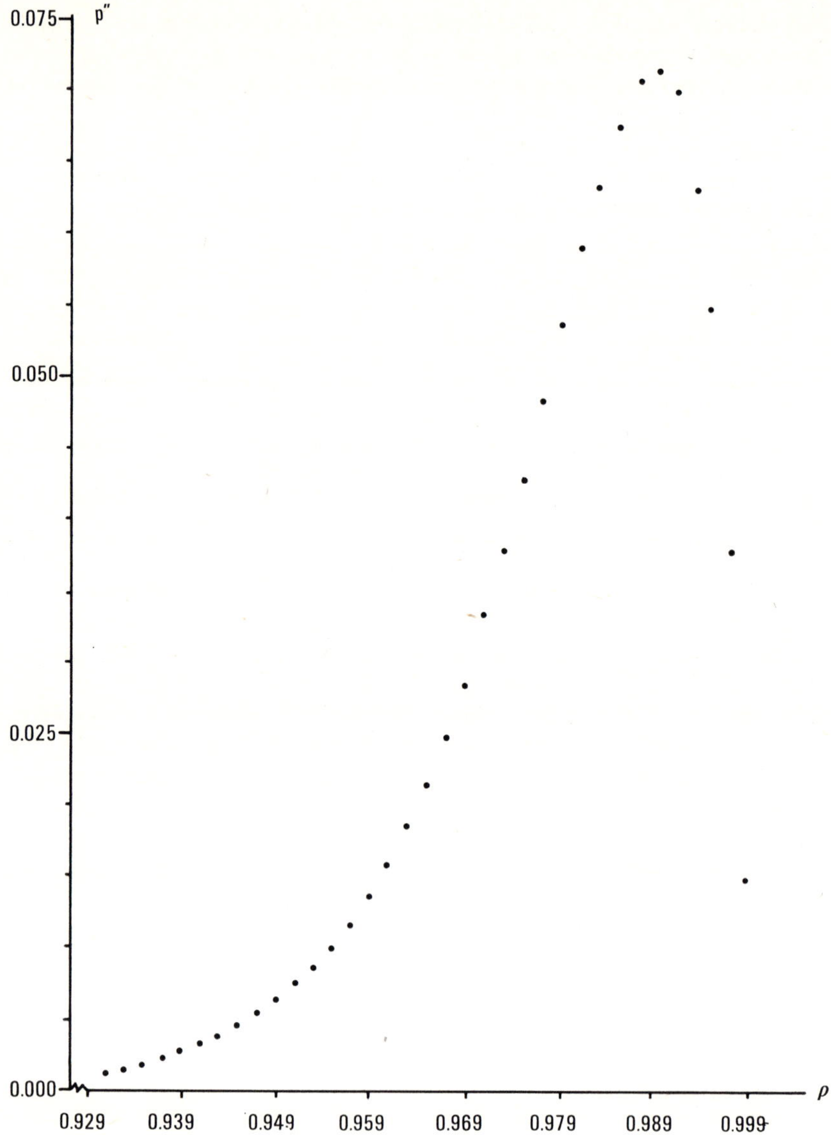

Fig. 2. Posterior probabilities for alternative values of ρ for the spirits example with observed data

results with earlier work in this field we refer to the already mentioned article of Zellner and Tiao (1964). Under slightly different assumptions concerning the autocorrelated disturbances, they derived a continuous posterior distribution of ρ given the sample z. We proceed to derive such a posterior distribution along their lines, but with the assumptions about the disturbances as presented in Section 4.3. Under these assumptions the kernel of the likelihood function is given by

$$(5.1) \qquad l(z|\beta, h, \rho) \propto |\Omega|^{-\frac{1}{2}} h^{\frac{1}{2}n} \exp\left[-\tfrac{1}{2}h(z-W\beta)^t \Omega^{-1}(z-W\beta)\right]$$

Accepting the locally uniform and independent prior densities

$$(5.2) \qquad D'(\beta) \propto k_1; \qquad D'(h) \propto h^{-1}; \qquad D'(\rho) \propto k_2$$

the following joint posterior density can be obtained

$$(5.3) \qquad D''(\beta, h, \rho|z) \propto |\Omega|^{-\frac{1}{2}} h^{\frac{1}{2}n-1} \exp\left[-\tfrac{1}{2}h\{\lambda v+(b-\beta)^t N(b-\beta)\}\right]$$

see (2.5). Upon integrating out the scale parameter h from (5.3), we obtain the following posterior density

$$(5.4) \qquad D''(\beta, \rho|z) \propto |\Omega|^{-\frac{1}{2}}\left[\tfrac{1}{2}\{\lambda v+(b-\beta)^t W^t \Omega^{-1} W(b-\beta)\}\right]^{-\frac{1}{2}n}$$

The marginal posterior density of ρ is

$$(5.5) \qquad D''(\rho|z) \propto |\Omega|^{-\frac{1}{2}} v^{-\frac{1}{2}\lambda} |W^t \Omega^{-1} W|^{-\frac{1}{2}}$$

where $\lambda = n-r$; see (3.11) and (3.12) for the dependence of Ω on ρ. We conclude that the ratio of posterior probabilities to alternative values of ρ (see (3.14)) is equal to the ratio of the corresponding values of the density (5.5).

The procedure of posterior probabilities can be applied to cases where the set of alternatives does not permit the use of a density function. We have seen, however, that if such a prior density exists, a straightforward Bayesian posterior distribution and posterior probabilities give identical results. In other words, the procedure of posterior probabilities is in a sense a more general approach than the existing ones mentioned above, but not essentially different.

4.6. THE NATURAL-CONJUGATE APPROACH TO NONSPHERICAL DISTURBANCES

In this section we reconsider the cases of alternative linear models with identical systematic parts and different assumptions concerning the disturbances for the case of natural-conjugate prior distributions. Let us assume that a Normal-gamma prior distribution is available for one of the models we consider. For simplicity, but without loss of generality, we assume that we have this prior distribution for our 'basic' model (see (2.1)), say model 1, with normally distributed disturbances η_1, such that $E\eta_1 = 0$ and $E\eta_1\eta_1^t = h_1^{-1}\Omega_1$. The problem we should like to solve is that of specification of the other Normal-gamma prior distributions. For this purpose we have two points in mind. First, all information from our known prior distribution must be used. Second, we have to avoid inconsistencies.

We start with the marginal prior distributions of β. As already stated in Section 4.2, we believe that our prior knowledge about β is not influenced by different assumptions concerning the disturbances. This implies that we accept equal marginal prior distributions on β for model 1, and, say, model 2, where model 2 corresponds to some arbitrary Ω_2. This, in turn, implies the following equalities for the parameters (see (2.4.9))

(6.1a) $\quad b_1' = b_2' = b'$

(6.1b) $\quad N_1'/v_1' = N_2'/v_2'$

(6.1c) $\quad \lambda_1' = \lambda_2' = \lambda'$

It follows from (6.1b) that neither N_2' nor v_2' is yet known. To obtain the values of these parameters we present the conditional distributions of η_1 and η_2, respectively. Both densities are normal with mean zero;

(6.2) $\quad f_N(\eta_i|h_i) = (2\pi)^{-\frac{1}{2}n} h_i^{\frac{1}{2}n} |\Omega_i|^{-\frac{1}{2}} \exp\{-\tfrac{1}{2}h_i\eta_i^t\Omega_i^{-1}\eta_i\}$

for $i = 1, 2$. We try to find a value of the ratio h_1/h_2 such that we are 'equally informative' in the sense that we require that our prior distributions of h_1 and h_2 express the same 'amount of information'. For this purpose we make use of information theory,[1] where the concept of average or expected information content is defined. The expected information of a distribution is often called the entropy of that distribution. For the density (6.2) the entropy is defined as

(6.3) $$-\int_{-\infty}^{\infty} \cdots \int_{-\infty}^{\infty} f(\eta_i|h_i) \log f_N(\eta_i|h_i)\, d\eta_i$$

which is identical to the mathematical expectation of

(6.4) $\quad \tfrac{1}{2}n \log 2\pi - \tfrac{1}{2}n \log h_i + \tfrac{1}{2} \log |\Omega_i| + \tfrac{1}{2}h_i\eta_i^t\Omega_i^{-1}\eta_i$

Only the last element of (6.4) is dependent on η_i. Its mathematical expectation is

(6.5) $\quad E\tfrac{1}{2}h_i\eta_i^t\Omega_i^{-1}\eta_i = \tfrac{1}{2}h_i E\,\mathrm{tr}\,[\Omega_i^{-1}\eta_i\eta_i^t] = \tfrac{1}{2}h_i\,\mathrm{tr}\,[\Omega_i^{-1}h_i^{-1}\Omega_i] = \tfrac{1}{2}n$

1. See, e.g., GOLDMAN (1955), Chapter IV.

where tr stands for the trace of a matrix,[1] and use has been made of the property tr $BA = \text{tr } AB$ if both products are defined. Combining (6.4) and (6.5) we find that (6.3) is equal to

$$(6.6) \qquad \tfrac{1}{2}n \log 2\pi - \tfrac{1}{2}n \log h_i + \tfrac{1}{2} \log |\Omega_i| + \tfrac{1}{2}n$$

so that equal entropies for $i = 1, 2$ are obtained if the following requirement is satisfied

$$(6.7) \qquad h_1^n/|\Omega_1| = h_2^n/|\Omega_2|$$

Now, we propose the following approach. If we start with the assumptions $E\eta_i\eta_i^t = h_i^{-1}\Omega_i^*$, where $\Omega_i^* = |\Omega_i|^{-1/n}\Omega_i$, implying $|\Omega_i^*| = 1$, then the original variance matrix of the disturbances is scaled such that for $h_1 = h_2$ equal entropies on the densities of these disturbances are obtained. Then equal prior distributions on h_1 and h_2 can be specified, so that $v_1' = v_2' = v'$. This, in turn (see (6.1 b)), implies $N_1' = N_2' = N'$, so that we conclude that identical Normal-gamma prior distributions for both alternative models can be used, provided that the variance matrices of the disturbances are scaled such that their determinants are equal.

One of the implications of the approach mentioned here is that the marginal prior densities of ε_1 and ε_2 (see (2.4) and Section 2.5) are equal. It also follows that the tops of the marginal densities of η_1 and η_2 have the same height; compare Section 2.5.

Combining the previous results, we find that the posterior probabilities as defined in (3.4.7) can be written as

$$(6.8) \qquad \frac{p_1''}{p_2''} = \frac{p_1'|N_2''|^{\frac{1}{2}}(\frac{1}{2}\lambda_2''v_2'')^{\frac{1}{2}\lambda_2''}(\frac{1}{2}\lambda_1''-1)!|J_1|}{p_2'|N_1''|^{\frac{1}{2}}(\frac{1}{2}\lambda_1''v_1'')^{\frac{1}{2}\lambda_1''}(\frac{1}{2}\lambda_2''-1)!|J_2|}$$

where b, v, N, and λ of (2.5) are substituted in (3.4.6) to obtain b'', v'', N'', and λ''. Some more simplifications can be introduced. By replacing (for $i = 1, 2$) N_i by $W^t(\Omega_i^*)^{-1}W$, λ_i'' by $\lambda' + n$, and $|J_i| = |\Omega_i^*|^{-\frac{1}{2}}$ by 1, we find

$$(6.9) \qquad \frac{p_1''}{p_2''} = \frac{p_1'|N' + W^t(\Omega_2^*)^{-1}W|^{\frac{1}{2}}(v_2'')^{\frac{1}{2}(\lambda'+n)}}{p_2'|N' + W^t(\Omega_1^*)^{-1}W|^{\frac{1}{2}}(v_1'')^{\frac{1}{2}(\lambda'+n)}}$$

It is interesting to investigate the behavior of (6.9) if the limiting approach is applied as discussed in Section 2.7. In other words, what happens if $p_1' = p_2'$ and N' and λ' tend to zero. Using the results as obtained in

1. The trace of a square matrix is defined as the sum of the elements on the main diagonal of that matrix.

Section 2.7 we find that the expression (6.9) tends to

$$(6.10)\qquad \frac{p_1''}{p_2''} = \frac{|W^t(\Omega_2^*)^{-1}W|^{\frac{1}{2}}(v_2^*)^{\frac{1}{2}n}}{|W^t(\Omega_1^*)^{-1}W|^{\frac{1}{2}}(v_1^*)^{\frac{1}{2}n}}$$

where v_i^* $(i=1, 2)$ is computed on the base of Ω_i^* by means of (2.5). It is also easy to verify that the uniform approach (compare (2.7)) for this case can be written as

$$(6.11)\qquad \frac{p_1''}{p_2''} = \frac{|W^t(\Omega_2^*)^{-1}W|^{\frac{1}{2}}(v_2^*)^{\frac{1}{2}(n-r)}}{|W^t(\Omega_1^*)^{-1}W|^{\frac{1}{2}}(v_1^*)^{\frac{1}{2}(n-r)}}$$

so that we conclude that the difference of (6.10) and (6.11) depends only on the different number of degrees of freedom. This result is of the same type as the result discussed in Section 2.7.

4.7. SENSITIVITY ANALYSIS

This section deals with some numerical sensitivity analysis on (6.9). It is clear that the choice of the values p_i' $(i=1, ..., k)$ and the values for the parameters b', v', N', and λ' influences the posterior probabilities and that this influence depends on the size of the sample. An analytical treatment of such a sensitivity analysis is too complicated to be of much interest. Therefore, we confine ourselves to a numerical illustration. For this illustration we use the first 17 observations of the spirits example already mentioned, and we compare two models which differ only in their autocorrelation parameter ρ, viz., $\rho_1=0.8$ and $\rho_2=0.9$. Concerning the parameters of the prior distribution the following remarks have to be made. Throughout the computations we assume our prior knowledge about the elements of β to be uncorrelated, implying that N' is a diagonal matrix. Further, we assume the median of the constant term to be zero with upper quartile M and lower quartile $-M$, and compute the posterior probabilities for $M\to\infty$ to express our lack of prior information concerning this term. For the medians of the income and price elasticities we take the values 1, 0.9, 0.8, and -1, -0.9, -0.8, respectively. For the upper quartiles we add 0.05, 0.1, 0.2 to each of these values. The lower quartiles follow by the symmetry of the Student distribution. For v' we take the values 0.5, 1, 2, 4, 8, 16, 32, and 64 times 10^{-4}. For λ' we choose the rather small values 1, 2, 3, 4, and 5, since the differences for larger values seem to be hardly noticeable if one looks at the figures in a table of cumulative Student distributions for $\lambda' > 5$. Finally, we accept $p_1' = p_2'$. The reader who has other prior ideas may easily adapt the results.

The reader will accept that it is nearly impossible to tabulate all the results of the computations described. We may even say that a complete presentation of the results is hardly interesting since these results are rather similar for the cases we consider. So, one of them will be presented here, others will only briefly be mentioned. First of all, in Table 4.2, one can find the upper bounds of 50 and 95 per cent Bayesian confidence intervals of the disturbances for specified values of λ' and v'. On the base of these figures one could exclude those combinations of λ' and v' which certainly do not correspond to one's prior beliefs. Secondly, in Table 4.3, we present an example of the posterior probabilities obtained for a case which is rather representative for all the other cases. The prior medians are 0, 1, and -1, respectively, the upper quartiles of the income and price elasticities are 1.2 and -0.8, and the 97.5 per cent fractiles for the income elasticity[1] are 3.451, 2.054, 1.831, 1.749, and 1.707 for λ' equal to 1, 2, 3, 4, and 5, respectively. Considering all results a first conclusion is that in all cases the same model is preferred. Further, for all the cases we considered, the largest change in probability due to changes in v' was obtained for $\lambda' = 5$, viz., from 0.78 to 0.65. But, one should not forget that the values of v' are varying from 0.00005 to 0.0064, which is rather impressive. For the smaller values of v' the probabilities do not change very much with changing λ', and even in the case of large values of v' these changes are small in comparison with the changes for $\lambda' = 5$ and changing v'. If one restricts the region of accepted λ' and v' values, by investigating the results of Table 4.2, for instance, then one might often conclude *a fortiori* that in this comparison problem the posterior probabilities are not very sensitive to these prior parameters. Fixing some values for λ' and v', and varying prior medians and quartiles, we found that the changes of the posterior probabilities are almost negligible for the varying quartiles, but for the varying medians these influences are of the same order of magnitude as changing the λ'.

Here again one might conclude that in the case in question the posterior probabilities are not very sensitive to the prior parameters. This, however, need not be a general conclusion, and we still advise anyone who uses prior information in a particular case, to carry out this type of sensitivity analysis.

1. For the price elasticity the corresponding fractiles are obtained by subtracting 2 from each of these numbers.

TABLE 4.2. UPPER BOUNDS OF 50 AND 95 PER CENT CONFIDENCE INTERVALS OF THE DISTURBANCES FOR ALTERNATIVE VALUES OF λ' AND v'

λ'	*1*	*2*	*3*	*4*	*5*
v'			50%		
0.00005	0.007	0.005	0.005	0.005	0.005
0.0001	0.010	0.008	0.007	0.007	0.007
0.0002	0.014	0.011	0.010	0.010	0.010
0.0004	0.020	0.016	0.015	0.014	0.014
0.0008	0.028	0.023	0.021	0.020	0.020
0.0016	0.040	0.032	0.030	0.029	0.029
0.0032	0.056	0.046	0.043	0.041	0.041
0.0064	0.080	0.065	0.061	0.059	0.058
			95%		
0.00005	0.089	0.030	0.022	0.019	0.018
0.0001	0.127	0.043	0.031	0.027	0.025
0.0002	0.179	0.060	0.045	0.039	0.036
0.0004	0.254	0.086	0.063	0.055	0.051
0.0008	0.359	0.121	0.090	0.078	0.072
0.0016	0.508	0.172	0.127	0.111	0.102
0.0032	0.718	0.243	0.180	0.157	0.145
0.0064	1.016	0.344	0.254	0.222	0.205

TABLE 4.3. POSTERIOR PROBABILITIES[1] FOR MODELS WITH $\rho = 0.9$, AND PRIOR MEDIANS 1, -1, AND UPPER QUARTILES 1.2, -0.8, FOR INCOME AND PRICE ELASTICITIES, RESPECTIVELY

λ'	*1*	*2*	*3*	*4*	*5*
v'					
0.00005	0.729	0.735	0.741	0.743	0.748
0.0001	0.730	0.733	0.733	0.737	0.737
0.0002	0.730	0.728	0.730	0.728	0.730
0.0004	0.732	0.724	0.721	0.719	0.718
0.0008	0.731	0.719	0.710	0.700	0.700
0.0016	0.721	0.703	0.693	0.686	0.681
0.0032	0.703	0.683	0.672	0.665	0.661
0.0064	0.681	0.662	0.653	0.648	0.645

1. The posterior probabilities for $\rho = 0.8$ can easily be computed by subtracting the posterior probabilities for $\rho = 0.9$ from 1.

First differences

A model often applied in time-series analysis deals with first differences of the variables. If one assumes the presence of a linear trend, for instance, and introduces time explicitly as an explanatory variable into the linear model, then the first-difference procedure eliminates this trend. Leser (1968) argues that this may be justified on computational grounds or to facilitate the search for suitable additional variables. Johnston (1963) states that in the absence of any information about the autocorrelation parameter ρ, some econometricians have taken it as approximately unity, so that first differences produces variables which are in a form suitable for the direct application of simple least squares. For a practical application of this approach reference is made to Stone (1954). Finally, the first-difference approach is often used in systems of demand equations or econometric macromodels if one is not interested in explaining or predicting the levels of certain economic phenomena, but is rather interested in explaining or predicting the changes (or relative changes) of these phenomena.

Suppose one likes to compare a level model with the corresponding first-differences model. The level model, say model 1, is defined as $z = W\beta + \eta_1$ (compare (4.2.1)) and the first-differences model (model 2) as $y = X\beta + \varepsilon_2$, where $y = Gz$, $X = GW$, G is an $(n-1) \times n$ matrix defined as

$$\text{(B.1)} \qquad G = \begin{bmatrix} -1 & 1 & 0 & 0 & \dots & 0 & 0 \\ 0 & -1 & 1 & 0 & \dots & 0 & 0 \\ 0 & 0 & -1 & 1 & \dots & 0 & 0 \\ \dots & \dots & \dots & \dots & \dots & \dots & \dots \\ 0 & 0 & 0 & 0 & \dots & -1 & 1 \end{bmatrix}$$

and $\varepsilon_2^t = [\varepsilon_2(2), \dots, \varepsilon_2(n)]$ is normally distributed with mean zero and variance matrix $h_2^{-1} I$. The disturbances η_1 are also assumed to be normally distributed[1] with mean zero and variance matrix $h_1^{-1} I$. The matrix G is not a square matrix so that we are not able to transform the likelihood of y into the likelihood of z. This implies that comparison of both models is not straightforwardly possible.

1. If one likes to compare a first-differences model and the corresponding level model with an autocorrelation parameter close to one, one can easily extend the method proposed in this appendix by making use of the results obtained in Section 4.3.

A possible solution of this problem is to transform model 1 with the matrix G. Note that this implies that we drop the possibility of explaining the level of the first observation. If we assume that this transformation does not change the probability of model 1, the comparison problem of the models is reduced to comparing models with equal variables to be explained, equal systematic parts and different assumptions about the disturbances, since $G\eta_1 = \varepsilon_1$ with

(B.2) $$E\varepsilon_1 = 0 \qquad E\varepsilon_1 \varepsilon_1^t = EG\eta_1 \eta_1^t G^t = h_1^{-1} GG^t$$

where the matrix GG^t is equal to

(B.3) $$GG^t = \begin{bmatrix} 2 & -1 & 0 & 0 & \dots & 0 & 0 \\ -1 & 2 & -1 & 0 & \dots & 0 & 0 \\ 0 & -1 & 2 & -1 & \dots & 0 & 0 \\ \cdots & \cdots & \cdots & \cdots & \cdots & \cdots & \cdots \\ 0 & 0 & 0 & 0 & \dots & 2 & -1 \\ 0 & 0 & 0 & 0 & \dots & -1 & 2 \end{bmatrix}$$

So the problem has been reduced to that of comparing models with alternative assumptions about the disturbances, which was discussed in Chapter 4. For the sake of completeness we note that $|GG^t| = n$. This can easily be seen by first-row expansion of $|GG^t|$ yielding

(B.4) $$\det(n-1) = 2\det(n-2) - \det(n-3)$$

where $\det(n-1)$ denotes the determinant of the $(n-1)\times(n-1)$ matrix GG^t for all n.

Some remarks have to be made. In some cases the matrix X has a zero column vector. This occurs if there is a constant term in the level model. Then X has no full column rank and the matrix X^tX is singular, which is not admitted in the case of no prior information. This, however, is not a serious problem. We can delete this constant term before the transformation with the matrix G, since after the transformation it disappears anyway. Moreover, the loss of the first observation is then not noticeable in the degrees of freedom of the posterior distribution.

As a second remark we call attention to the assumption that the G transformation does not change the probability of model 1. This assumption is rather arbitrary and could be defended by stating that one believes *a priori* that deleting the possibility of explaining the first (level) observation has little or no influence on the features of the model. If one would not accept this assumption, one could introduce $y(1)$ as

$$\text{(B.5)} \qquad y(1) = z(1) - z(0) = \sum_{j=1}^{r_2} x_j(1)\beta_j + \varepsilon_2(1)$$

or, since no observation on $z(0)$ is available,

$$\text{(B.6)} \qquad z(1) = \sum_{j=1}^{r_1} w_j(1)\beta_j + \varepsilon_2(1) + \eta^*(0)$$

where use has been made of $z(0) = \sum_{j=1}^{r_1} w_j(0)\beta_j + \eta^*(0)$. This approach also needs some additional assumptions. To be precise, if one assumes that $\eta^*(0)$ is uncorrelated with $\varepsilon_2(t)$ $(t = 1, \ldots, n)$, and that $\eta^*(0)$ is normally distributed with mean zero and variance δh_2^{-1}, where δ is some known positive number, then the comparison problem can be solved. This can easily be seen by transforming the level model with an $n \times n$ matrix G^* defined as

$$\text{(B.7)} \qquad G^* = \begin{bmatrix} 1 & 0 \\ & G \end{bmatrix}$$

Then the problem of comparing a level model with the corresponding first-differences model is also reduced to comparing models with alternative assumptions about the disturbances. Note, however, that generally the value of δ is not known *a priori*. Then it is possible to compute for which values of δ model 1 or model 2 is preferred,[1] or to specify a subjective distribution of δ. Finally, this approach also implies that if we transform this model to a level model, the disturbances are heteroskedastic, and therefore not comparable with $\rho \to 1$ in the case of autocorrelation.

1. 'Preferred' is used in the sense of 'more probable' by THORNBER (1966) and GEISEL (1970). We shall often follow this usage.

5. Different sets of explanatory variables

5.1. INTRODUCTION

In this chapter we shall discuss the problem of comparing alternative linear models which have the same variable to be explained, but different matrices of explanatory variables. Throughout this chapter it is assumed that the disturbances of the models are independently and normally distributed.

Once we have specified the state space of the models to be considered, the problem arises how to assign the prior probabilities to the models and the prior distributions of the parameters, for the uniform as well as for the Normal-gamma case. We do not pay much attention to the Normal-gamma case, but mainly repeat some of the remarks already made in Section 3.7. If the investigator is able to specify values for $p'_i, b'_i, N'_i, \lambda'_i$, and v'_i $(i = 1, \ldots, k)$, he can apply (3.4.7) without difficulty. Since we believe that such a task is generally quite complicated, we advise anyone who uses this approach to carry out sensitivity analysis for acceptable variations of the prior parameters.

For an application of fitting prior information to a Normal-gamma distribution we already referred to Harkema and Kloek (1969). They compare alternative proxy-variables for explanatory variables which cannot be measured. In such a case one may often assume (as they do) that the prior parameters are independent of the way in which the explanatory variables are approximated. This implies that $b'_i = b'_j$, $N'_i = N'_j$, $\lambda'_i = \lambda'_j$, and $v'_i = v'_j$ for $i, j = 1, \ldots, k$, which simplifies the difficulties inherent in the Normal-gamma approach to a great extent.

5.2. THE CASE OF NO PRIOR INFORMATION

Let us now turn to the case where the investigator has very little information on β and h. Then, he may wish to consider the limiting behavior of (3.5.11) as α tends to infinity. For the case of comparing models with different X matrices, however, we believe that it is impossible to give a general rule for obtaining a unique limit of $\gamma'_{\alpha i}/\gamma'_{\alpha j}$. There are some particular cases for which a simple solution is available, for example, for a set of Koyck-type distributed-lag models, where the most probable lag parameter must be found. A more detailed discussion of these models can be found in Appendix C.

It is of some interest to investigate the invariance of (3.5.11) in the case where $X_j = X_i A$, where A is some known nonsingular matrix. This situation is quite unrealistic for practical purposes, since there is no essential difference between the models. It is, however, of theoretical interest to study the consequences of such a comparison problem. Suppose, we assign equal independent uniform prior densities[1] to β_i and $A\beta_j$, and equal uniform prior densities to $\log h_i$ and $\log h_j$, so that $\gamma'_{\alpha i}/\gamma'_{\alpha j} = |A|^{-1}$ for all α. It is easy to verify that this implies that $p''_{\alpha i}/p''_{\alpha j} = p'_i/p'_j$. We conclude, therefore, that the posterior probabilities are invariant for changes of the units of measurement (if A is diagonal) and that the ratio of posterior probabilities of two models is equal to the ratio of the prior probabilities if the explanatory variables of one model are linear combinations of the explanatory variables of the other model, provided that these combinations are obtained by means of a nonsingular linear transformation.

Let us return to the general problem: how to treat cases of different X matrices. We first discuss the already mentioned studies[2] of Thornber (1966) and Geisel (1970). Thornber adopts a procedure introduced by Jeffreys (1961). His 'invariant' prior density is based on the square root of the determinant of the so-called information matrix,[3] with elements

$$I_{ij}(\theta) = -E\{\partial^2 \log l(z|\theta)/\partial\theta_i \partial\theta_j\}$$

where the expectation has to be taken with respect to the conditional distribution of z, given θ. The resulting prior density has the following shape

$$\text{(2.1)} \qquad f'_\alpha(\beta_i, h_i) = \bar{\gamma}'_{\alpha i}|X'_i X_i|^{\frac{1}{2}} h^{\frac{1}{2}r_i - 1} \quad \text{if} \quad (\beta_i, h_i) \in B_\alpha$$
$$= 0 \qquad \text{otherwise}$$

and B_α is defined in (3.5.1). Then Thornber obtains general expressions for

1. Note that (unless A is a diagonal matrix) the distribution of β_j is uniform on a non-rectangular region.
2. The work of ZELLNER (1971) is very similar to that of GEISEL (1970), and will not be discussed explicitly.
3. See also KENDALL and STUART (1961), Section 17.39.

the posterior probabilities by assuming[1] (without comment) that $\bar{\gamma}'_{\alpha i} = \bar{\gamma}'_{\alpha j}$ for $i, j = 1, \ldots, k$. In the case of comparing models with the same number of explanatory variables, the same variable to be explained, and the same prior probabilities, his general expression reads as follows

$$(2.2) \qquad \frac{p''_i}{p''_j} = \frac{(\lambda_j v_j)^{\frac{1}{2}n}}{(\lambda_i v_i)^{\frac{1}{2}n}}$$

and he concludes that model i is preferred to model j when $\lambda_i v_i < \lambda_j v_j$. This means that the criterion of maximizing p''_i is equivalent to that of maximizing R^2. For the case of a different number of explanatory variables, $r_i \neq r_j$, Thornber obtains an 'adjusted' rule which is more complicated than (2.2). His conclusion is that this 'adjusted' rule chooses the simpler model (that is the model with the smallest number of explanatory variables) unless there is a fairly large reduction in the sum of the squared estimated disturbances if more explanatory variables have been introduced.

Note that by using the prior density (2.1) a marginal posterior distribution on β is generated, which is Student (or t) with n degrees of freedom independent of the dimension of β. Jeffreys (1961), who discussed the same phenomenon for a similar but simpler case, is not very happy with this property. He derives this prior density (2.1) for one location parameter λ (corresponding to a one-dimensional β) and a scale parameter σ, and concludes (see Jeffreys, p. 182): 'If the same method was applied to a joint distribution for several variables about independent true values, an extra factor $1/\sigma$ would appear for each. The index in the corresponding t distribution would always be $\frac{1}{2}(n+1)$ however many true values were estimated. This is unacceptable. In the usual situation in an estimation problem λ and σ are each capable of any value over a considerable range, and neither gives any appreciable information about the other'. Next, he continues to derive another prior density for the multivariate case, which we used in the Sections 2.6 and 3.5.

Geisel (1970) and Zellner (1971), on the other hand, start with proper natural-conjugate (Normal-gamma) prior distributions. They assume equal prior probabilities $p'_i = 1/k$ for $i = 1, \ldots, k$, and apply the method of decreasing prior information as discussed in Section 2.7, such that the sample information dominates the prior information. To obtain tractable results, they also assume that $\lambda'_i = \lambda'$, $v'_i = v$, and that

$$|N'_i|^{\frac{1}{2}}\ |N''_j|^{\frac{1}{2}} / |N'_j|^{\frac{1}{2}}\ |N''_i|^{\frac{1}{2}}$$

1. The 'invariant' prior density (2.1) gives neither a unique result for (3.5.11). Dagenais (1971) suggests to accept $\bar{\gamma}'_{\alpha i} = c r_i^{-\frac{1}{2}}$, and also passes the essential problem to assign a value to the limit of $\gamma'_{\alpha i}/\gamma'_{\alpha j}$.

tends to one for all $i, j=1, \ldots, k$. Then for $\lambda' \to 0$ and $N_i' \to 0$ $(i=1, \ldots, k)$ they obtain in the case of comparing models with the same number of explanatory variables and the same variable to be explained the same result as is obtained in (2.2), again justifying the use of R^2 in choosing from such a set. This result is easily verified if one carries out the above described approach on (3.4.7).

For the case of comparing models with a different number of explanatory variables Geisel (1970) introduces some additional assumptions such that his procedure yields the same result as the adjusted R^2. As Thornber, he also concludes that his assumptions imply a strong prior preference for the model with the smaller number of explanatory variables.

Further, it is easy to see that the natural-conjugate prior of Geisel's approach is proportional to $|X_i^t X_i|^{\frac{1}{2}} Th_i^{\frac{1}{2}r_i - 1}$ in the limiting case, so that ultimately there is no difference in the approaches of Thornber and Geisel.

The only advantage, in our opinion, of the procedures discussed, is that they produce a mechanical rule, which can be applied straightforwardly to practical problems. Concerning the practical application of (2.2) some remarks can be made. We carried out some simulation computations for the purpose of comparing two models containing the same number of explanatory variables and the same variable to be explained. We found that in most cases the correct model, that is the one we specified in advance, was preferred to the other. We also computed the posterior probabilities of the two models for different types of X matrices and we also interchanged the X matrices of the two models. The resulting posterior probabilities appeared to be strongly influenced by the choice of the X matrix of the correct specification.[1] For some X matrices the correct model always obtained the largest posterior probability. For others a large number of small posterior probabilities (below 0.5) were found for the correct model. Finally, corresponding to the results obtained in Section 3.6, we found a decreasing number of wrong decisions in the case of increasing sample sizes.

5.3. REVIEW OF THE MECHANICAL RULE

In comparing linear models with equal variables to be explained and different explanatory variables, the prior distributions as discussed in Section 5.2 all depended on the matrix of explanatory variables. These approaches, however, seem rather arbitrary to us. If, for example, one additional observation comes available, then our prior beliefs have to be changed. This gives untenable results. Suppose we carry out our sampling

1. This result is very similar to the results of KOERTS and ABRAHAMSE (1969), Section 8.4.

procedure in two phases. Let the total sample of n observations be partitioned such that the first part consists of the first n_1 observations and the second of the observations $n_1+1, \ldots, n$. Then the models can be rewritten as

$$(3.1) \qquad \begin{bmatrix} z_1 \\ \\ z_2 \end{bmatrix} = \begin{bmatrix} X_{i1} \\ \\ X_{i2} \end{bmatrix} \beta_i + \begin{bmatrix} \varepsilon_{i1} \\ \\ \varepsilon_{i2} \end{bmatrix} \qquad (i = 1, \ldots, k)$$

where the numbers 1 and 2 denote the two sampling phases. It is clear that in combining the likelihood of the first sample with a uniform prior density on β and log h, the following Normal-gamma posterior density is obtained

$$(3.2) \qquad D''(\beta_i, h_i|z_1) \propto h_i^{\frac{1}{2}n_1-1} \exp[-\tfrac{1}{2}h_i\{\lambda_{i1} v_{i1} + \\ + (\beta_i - b_{i1})^t X_{i1}^t X_{i1}(\beta_i - b_{i1})\}]$$

where b_{i1}, λ_{i1}, and v_{i1} are the first-sample analogues of b_i, λ_i, and v_i. Interpreting (3.2) as a Normal-gamma prior density for the second sample, the resulting posterior density reads as follows

$$(3.3) \qquad D''(\beta_i, h_i|z) \propto h_i^{\frac{1}{2}n-1} \exp[-\tfrac{1}{2}h_i\{\lambda_{i1} v_{i1} + \lambda_{i2} v_{i2} + \\ + (\beta_i - b_{i1})^t X_{i1}^t X_{i1}(\beta_i - b_{i1}) + (\beta_i - b_{i2})^t X_{i2}^t X_{i2}(\beta_i - b_{i2})\}]$$

where b_{i2}, λ_{i2}, and v_{i2} are based on the second part of the sample. This can be rewritten as

$$(3.4) \qquad D''(\beta_i, h_i|z) \propto h_i^{\frac{1}{2}n-1} \exp[-\tfrac{1}{2}h_i\{\lambda_i v_i + (\beta_i - b_i)^t X_i^t X_i(\beta_i - b_i)\}]$$

(compare Section 2.4) so that we find that this ultimate result is equal to the result of combining the whole sample with a uniform prior distribution.

We continue to discuss the consequences of this two-phase sampling approach for the computation of the posterior probabilities. Compare (3.5.11) to conclude that the ratio of posterior probabilities given the first sample is equal to

$$(3.5) \qquad \frac{p''_{i1}}{p''_{j1}} = \gamma_{ij} \frac{|X_{j1}^t X_{j1}|^{\frac{1}{2}} v_{j1}^{\frac{1}{2}\lambda_1}}{|X_{i1}^t X_{i1}|^{\frac{1}{2}} v_{i1}^{\frac{1}{2}\lambda_1}}$$

for $p'_i = p'_j$, $r_i = r_j$ (thus $\lambda_{i1} = \lambda_{j1} = \lambda_1$). Note that p''_{i1} now denotes the posterior probability of model i after the first n_1 observations and[1] $\gamma_{ij} = \lim_{\alpha \to \infty} \gamma'_{\alpha i}/\gamma'_{\alpha j}$. Next, we accept the posterior probabilities p''_{i1} $(i = 1, \ldots, k)$ and the posterior distributions given the first n_1 observations as the prior probabilities and prior distributions for the second part of the sample.

1. We recall that (3.5) can be applied only if the investigator is prepared to specify γ_{ij}.

Combining (3.4.7) and (3.5), yields the following expression for the ratio of posterior probabilities given the whole sample

$$(3.6)\qquad \frac{p_i''}{p_j''} = \frac{p_{i1}''|X_{i1}^t X_{i1}|^{\frac{1}{2}} |X_j^t X_j|^{\frac{1}{2}} v_{i1}^{\frac{1}{2}\lambda_1} v_j^{\frac{1}{2}\lambda}}{p_{j1}''|X_{j1}^t X_{j1}|^{\frac{1}{2}} |X_i^t X_i|^{\frac{1}{2}} v_{j1}^{\frac{1}{2}\lambda_1} v_i^{\frac{1}{2}\lambda}}$$

which can be rewritten (compare (3.5)) as

$$(3.7)\qquad \frac{p_i''}{p_j''} = \gamma_{ij} \frac{|X_j^t X_j|^{\frac{1}{2}} v_j^{\frac{1}{2}\lambda}}{|X_i^t X_i|^{\frac{1}{2}} v_i^{\frac{1}{2}\lambda}}$$

This result is completely identical to the result that one obtains by applying (3.5.11) to the whole sample at once. This property satisfies an intuitive requirement[1] that a two-phase sampling approach does not influence the ultimate result, since equal sample information has to give equal posterior results.

The approach (2.2), however, does not possess this attractive property. If we apply the procedure discussed in (3.5), (3.6), and (3.7) to the approach (2.2) we find

$$(3.8)\qquad \frac{p_{i1}''}{p_{j1}''} = \frac{v_{j1}^{\frac{1}{2}n_1}}{v_{i1}^{\frac{1}{2}n_1}}$$

and

$$(3.9)\qquad \frac{p_i''}{p_j''} = \frac{|X_{i1}^t X_{i1}|^{\frac{1}{2}} |X_j^t X_j|^{\frac{1}{2}} v_j^{\frac{1}{2}n}}{|X_{j1}^t X_{j1}|^{\frac{1}{2}} |X_i^t X_i|^{\frac{1}{2}} v_i^{\frac{1}{2}n}}$$

which is in general[2] not equal to $v_j^{\frac{1}{2}n}/v_i^{\frac{1}{2}n}$.

Given this consideration, we believe that any general assumption concerning relations between prior parameters and X matrices, has to be called an arbitrary assumption. Although in practical situations research workers may apply (2.2) and be satisfied with the results, we state that in our opinion the lack of any logical foundation weighs so heavily that we would not advise the application of this mechanical rule.

1. Of course, this requirement belongs to the 'objectivistic' philosophy. Subjectivists need not accept an intermediate posterior distribution as their prior for later periods, since they may disagree with it.
2. In the case where all circumstances which influence some experiment can be fixed by the investigator, e.g., in biological or agricultural research (compare Section 1.2), he can repeat the experiment several times with the same X matrix, and (3.9) can be equal to $v_j^{\frac{1}{2}n}/v_i^{\frac{1}{2}n}$. In economics, however, it is usually impossible to repeat the experiment with the same X matrix.

A possible way out is to make use of the tool of subjective knowledge. It will be clear that such an approach does not generate a mechanical rule (or procedure) and will not always yield better decisions than the mechanical rule we considered. One is, however, conscious of the arbitrariness of assigning a value to the limit of $\gamma'_{\alpha i}/\gamma'_{\alpha j}$.

It is interesting to discuss some of the possibilities according to an increasing number of parameters that have to be specified in a subjective way.

(i) A straightforward fixation of the value $\gamma'_{\alpha i}/\gamma'_{\alpha j}$ for all α. The possibility of obtaining a reasonable approach strongly depends on the alternative specifications of the models which have to be considered. We already mentioned the rather natural solutions in the case that the models differ by the assumptions concerning the disturbances only; see Chapter 4. For the case of different X matrices we believe that if the models are constructed such that the explanatory variables do not differ too much, or are related in some way to each other, it may be possible to find a value of $\gamma'_{\alpha i}/\gamma'_{\alpha j}$ or an interval to which $\gamma'_{\alpha i}/\gamma'_{\alpha j}$ belongs in the opinion of the investigator. If such an interval is defined, he can investigate whether the same model is always preferred to the other models. It will be clear that the above mentioned approach does not generate exact posterior probabilities if one is not sure about the value of $\gamma'_{\alpha i}/\gamma'_{\alpha j}$, but that in certain cases it is possible to indicate the most probable model.

(ii) A two-phase sampling approach. The first n_1 sample observations can be used to analyse extensively the alternative models and to specify the prior distributions of the parameters before the second sample is observed. These prior distributions can be obtained by combining the likelihoods of the first sample with uniform prior distributions. It is possible that if one observes these results, one does not agree with all signs and magnitudes of the medians of β. Then it is possible to introduce some changes to these distributions according to this now revealed prior knowledge, which was latent before. Finally, the second sample can be used to investigate whether this sample confirms the investigator's prior ideas. A more detailed discussion and a numerical example of this approach can be found in Chapter 6. This chapter also deals with the comparison of a specification which is linear in the logarithms and a specification which is linear in the variables themselves.

(iii) The straightforward assessment of prior probabilities and natural-conjugate prior distributions as has already been suggested in Section 5.1. Note that this approach is in a sense an extreme case of the method suggested in (ii), to be more precise, in the sense that no sample evidence has been used to obtain prior probabilities and prior distributions.

Distributed lags

In several econometric studies it is assumed that the effect of a change in an explanatory variable on the variable to be explained is not immediate, but is subject to a distributed lag over time. Such a model is often written in the form

$$(C.1) \qquad y(t) = \beta_0 + \beta_1 x(t) + \beta_2 x(t-1) + \ldots + u(t)$$

Estimation of the parameters presents obvious difficulties since our sample will be of finite size, so the infinite set of lagged regressors must be cut off at some point. Even then these successive lagged economic variables are often highly correlated, so that multicollinearity has to be expected. Several restrictive assumptions for the parameters of (C.1) based on considerations of simplicity are developed, but will not all be discussed in this book. A well-known *a priori* restriction on the parameters, originally suggested by Irving Fisher, extended by Koyck (1954), and reconsidered by Nerlove (1958), is to assume that the sequence $\{\beta_i\}_{i=1}^{\infty}$ is, from a certain point onwards, a geometric sequence. We consider the simplest case, in which the sequence $\{\beta_i\}_{i=1}^{\infty}$ is geometric right from the beginning. Then, we can replace (C.1) by

$$(C.2) \qquad y(t) = \beta_0 + \beta_1 x(t) + \delta\beta_1 x(t-1) + \delta^2 \beta_1 x(t-2) + \ldots + u(t)$$

where $0 < \delta < 1$. Shifting equation (C.2) one period backward, multiplying it by δ and subtracting this result from (C.2), we obtain

$$(C.3) \qquad y(t) - \delta y(t-1) = (1-\delta)\beta_0 + \beta_1 x(t) + \{u(t) - \delta u(t-1)\}$$

or

$$(C.4) \qquad y(t) - \delta y(t-1) = (1-\delta)\{\beta_0 + \beta_1^* x(t)\} + \{u(t) - \delta u(t-1)\}$$

where $\beta_1^* = \beta_1/(1-\delta)$. Expression (C.4) can be written as

$$(C.5) \qquad y(t) = \delta y(t-1) + (1-\delta)\{\beta_0 + \beta_1^* x(t)\} + \{u(t) - \delta u(t-1)\}$$

Suppose that $\beta_0 + \beta_1^* x(t)$ is the equilibrium level[1] of $y(t)$ given $x(t)$, then (C.5) represents the partial adaptation of the actual level to this equilibrium level. Considering (C.4), it should be noted, that the disturbance is

1. Note that MALINVAUD (1966), p. 477 assumes that this equilibrium relation contains a random disturbance. Additional assumptions are required in order to obtain an identifiable structure if we accept this assumption in our approach.

correlated with the explanatory variable $y(t-1)$, unless $u(t)$ obeys the Markov scheme

$$(C.6) \qquad u(t) = \rho u(t-1) + \eta(t)$$

with $\rho = \delta$ and η is uncorrelated over time and independent of the explanatory variables $x(t)$ and $y(t-1)$. The assumption that $\rho = \delta$ is simple, but not always realistic; so we drop this assumption to obtain a more general approach.

Several consistent two-step point estimation procedures of the parameters ρ and δ are developed in the previously cited works of Koyck and Nerlove. They also considered extensions for handling more explanatory variables. Further work in this field has been done by, e.g., Klein (1958), Solow (1960), Griliches (1961), Liviatan (1963), Almon (1965), Jorgenson (1966), Dhrymes (1969), Gupta (1969), and Heien (1969).

In Bayesian statistics this estimation problem can be handled by assuming the disturbances to be normally distributed and assigning uniform prior distributions on β_0, β_1, $\log h$, δ and ρ. Sometimes, if some prior information on β_1 is available (and the constant term is deleted), a Beta distribution is assigned. Combining the prior distribution with the likelihood of the sample yields a posterior distribution of these parameters. Then marginal and conditional posterior distributions of δ and ρ can be derived. See, e.g., Zellner and Park (1965), Thornber (1966), Zellner and Geisel (1968), and Geisel (1970).

In the context of the present study it is obvious that we investigate the application of the procedure of posterior probabilities to distributed-lag models. We consider a finite number of pairs (δ, ρ), say k_1 values of δ and k_2 values of ρ, for the purpose of constructing $k = k_1 k_2$ models and confine ourselves to the case of uniform prior distributions. We redefine (C.4) as

$$(C.7) \qquad w(t) \equiv y(t) - \delta y(t-1) = (1-\delta)\{\beta_0 + \beta_1^* x(t)\} + \{u(t) - \delta u(t-1)\}$$

for $t = 1, \ldots, n$, and assume that n observations are available on $y(t)$ and $x(t)$, all $x(t)$ are non-stochastic, and that the $u(t)$ are normally distributed with constant variances following a first-order Markov-scheme as discussed in Section 4.3. We consider the vector w of (C.7) to be of dimension n, although $y(0)$ is not observable. This problem can be handled rather easily, as will be shown. Therefore we write the left-hand side of (C.7) in matrix notation as follows

$$(C.8) \qquad w = Gy - \delta y(0) \mathrm{e}_1$$

where e_1 is the n-dimensional unit vector defined by $\mathrm{e}_1^t = (1\ 0\ 0\ \ldots\ 0)$, and

$$(C.9) \quad G = \begin{bmatrix} 1 & 0 & 0 & \dots & 0 & 0 \\ -\delta & 1 & 0 & \dots & 0 & 0 \\ 0 & -\delta & 1 & \dots & 0 & 0 \\ \cdots & \cdots & \cdots & \cdots & \cdots & \cdots \\ 0 & 0 & 0 & \dots & -\delta & 1 \end{bmatrix}$$

and

$$G^{-1} = \begin{bmatrix} 1 & 0 & 0 & \dots & 0 & 0 \\ \delta & 1 & 0 & \dots & 0 & 0 \\ \delta^2 & \delta & 1 & \dots & 0 & 0 \\ \cdots & \cdots & \cdots & \cdots & \cdots & \cdots \\ \delta^{n-1} & \delta^{n-2} & \delta^{n-3} & \dots & \delta & 1 \end{bmatrix}$$

Now, (C.8) can be rewritten as

$$(C.10) \quad y = G^{-1}w + \delta y(0)\, G^{-1}\mathrm{e}_1$$

which, in turn, can be combined with (C.7), so that we obtain

$$(C.11) \quad y = (1-\delta)G^{-1} \begin{bmatrix} 1 & x(1) \\ 1 & x(2) \\ \vdots & \vdots \\ 1 & x(n) \end{bmatrix} \begin{bmatrix} \beta_0 \\ \beta_1^* \end{bmatrix} + \\ + \delta y(0)\, G^{-1}\mathrm{e}_1 + G^{-1} \begin{bmatrix} u(1)-\delta u(0) \\ u(2)-\delta u(1) \\ \vdots \\ u(n)-\delta u(n-1) \end{bmatrix}$$

This yields

$$(C.12) \quad y = (1-\delta)\, G^{-1} \begin{bmatrix} 1 & x(1) \\ 1 & x(2) \\ \vdots & \vdots \\ 1 & x(n) \end{bmatrix} \begin{bmatrix} \beta_0 \\ \beta_1^* \end{bmatrix} + \\ + \delta G^{-1} \begin{bmatrix} y(0)-u(0) \\ 0 \\ \vdots \\ 0 \end{bmatrix} + \begin{bmatrix} u(1) \\ u(2) \\ \vdots \\ u(n) \end{bmatrix}$$

The term $y(0)-u(0)$ is, given the sequence $\{\beta_i\}_{i=1}^{\infty}$, non-stochastic and unknown. Therefore, it is possible to treat it as a parameter to be estimated in the same way as β_0 and β_1^*. This enables us to rewrite (C.12) as follows

(C.13) $$y = (1-\delta)G^{-1}[\{\delta/(1-\delta)\}\,\mathrm{e}_1 \quad \iota \quad x]\begin{bmatrix} y(0)-u(0) \\ \beta_0 \\ \beta_1^* \end{bmatrix} + u$$

where ι is an n-dimensional vector with all elements equal to one. Let us replace the matrix of the systematic part by the matrix A, giving

(C.14) $$y = A\beta + u$$

where

(C.15) $$A = (1-\delta)G^{-1}[\{\delta/(1-\delta)\}\mathrm{e}_1 \quad \iota \quad x]$$

and

(C.16) $$\beta^t = [y(0)-u(0) \quad \beta_0 \quad \beta_1^*]$$

The likelihood of (C.14) reads as follows

(C.17) $$l(y|\beta, h, \rho, \delta) = (2\pi)^{-\frac{1}{2}n} h^{\frac{1}{2}n} |\Omega^{-1}|^{\frac{1}{2}} \exp\{-\tfrac{1}{2}h(y-A\beta)^t \Omega^{-1}(y-A\beta)\}$$

where $h^{-1}\Omega$ is the variance matrix of the disturbances u; see Section 4.3.

The construction of the uniform prior distributions is rather easy. Concerning the disturbances it follows from Section 4.2 that we can assign identical uniform prior distributions on the logarithm of the precisions of the alternative models. Next, we consider the β vector as defined in (C.16) and start with $y(0)-u(0)$, the systematic part of (C.1) in the time period zero. In our opinion there is no reason to believe that there is a different amount of information concerning this term for the alternative models, and we assign the same prior distribution to $y(0)-u(0)$ for every model. Further, all models are assumed to have the same long term equilibrium level $\beta_0+\beta_1^* x(t)$ of $y(t)$, given $x(t)$. So, we also accept identical prior distributions for β_0 as well as β_1^* for the alternative models. Then, $\gamma'_{\alpha i}/\gamma'_{\alpha j} = 1$ for all α, and if we accept equal prior probabilities, the formula[1] of posterior probabilities (3.5.11) can be written as

(C.18) $$\frac{p''_i}{p''_j} = \frac{|\Omega_j|^{\frac{1}{2}}|A_j^t \Omega_j^{-1} A_j|^{\frac{1}{2}}(\frac{1}{2}\lambda v_j)^{\frac{1}{2}\lambda}}{|\Omega_i|^{\frac{1}{2}}|A_i^t \Omega_i^{-1} A_i|^{\frac{1}{2}}(\frac{1}{2}\lambda v_i)^{\frac{1}{2}\lambda}}$$

1. Compare (4.2.2)–(4.2.7). The simplified expression for $|\Omega|$ obtained in (4.3.14) has not been substituted, since (C.18) holds for arbitrary Ω.

where i and j denote two different models with alternative combinations of δ and ρ.

A brief concluding remark can be made. Accepting the locally uniform and independent prior densities

$$\text{(C.19)} \quad D'(\beta) \propto c_1\,; \quad D'(h) \propto h^{-1}\,; \quad D'(\rho) \propto c_2\,; \quad D'(\delta) \propto c_3$$

for $-\infty < \beta < \infty$, $h > 0$, $-1 < \rho < 1$, and $0 < \delta < 1$, and combining this prior with the likelihood function as defined in (C.17), the following marginal posterior density of ρ and δ can be derived

$$\text{(C.20)} \quad D''(\rho, \delta|y) \propto |\Omega|^{-\frac{1}{2}} v^{-\frac{1}{2}\lambda} |A^t \Omega^{-1} A|^{-\frac{1}{2}}$$

so that we conclude that the ratio of posterior probabilities of alternative combinations of ρ and δ is equal to the ratio of the corresponding values of the density (C.20). For a similar result we refer to Section 4.5.

Note that in this appendix we confined ourselves to an infinite lag distribution. It is, however, possible that economic data may be generated by finite lag distributions; see, e.g., Almon (1965) and Heien (1969). It is also possible to drop the assumption that $\{\beta_i\}_{i=1}^{\infty}$ is a geometric sequence and to accept another distribution, for example, a Pascal distribution; see Solow (1960). Several of the other approaches can be attacked in a straightforward manner by means of posterior probabilities; this, however, is outside the scope of the present study.

6. Some experiments with alternative model specifications

6.1. INTRODUCTION

In this chapter we shall present a numerical illustration of the comparison of models with different explanatory variables and the same dependent variable as well as different, but functionally related, dependent variables. We use again the spirits example (see Section 4.4) and restrict ourselves to first-differences models to avoid the large amount of autocorrelation found in Section 4.4. We emphasize that we give a numerical illustration of the problems discussed in Chapters 3 and 5 and a tentative solution for comparing this kind of models, without claiming that we arrive at the model which provides the best explanation for the consumption of spirits in the United Kingdom from 1870 to 1938. We start to construct twelve alternative linear specifications, again without pretending to be exhaustive.[1] Certainly, critical readers can find other acceptable or more sophisticated models. Neither do we pretend that each of the models we consider satisfies all theoretical requirements that have been proposed in the literature.[2] Systems of demand equations, for instance, are hardly considered. We make use of the indirect addilog model (or expenditure allocation model), but for reasons of simplicity we neither consider the Rotterdam differential demand model of Theil and Barten, nor Stone's linear expenditure demand model.

The procedure is as follows. We have a sample of 69 observations, which is reduced to 68 observations since all models are specified in first differences. We divide this sample into two parts. This can be done in a great many ways, but we confine ourselves to the case where the first sample contains the observations 1, ..., 34 and the second sample the observations 35, ..., 68. Within each model the first sample is used to obtain a Normal-gamma posterior distribution by means of a uniform prior distribution.

1. Note that the choice of such a set of models *always* involves decisions taken on subjective grounds.
2. See, e.g., SOMERMEYER (1967).

This posterior distribution is then available as the prior distribution which can be combined with the likelihood of the second sample. This prior distribution yields the values of b_i', N_i', λ_i', and v_i' (if model i is considered), but not p_i' which is equal to the posterior probability of model i after the first sample, since (3.5.11) can not be applied because of the unknown value of $\gamma_{\alpha i}'/\gamma_{\alpha j}'$ $(i, j = 1, \ldots, k;\ i \neq j)$. The two-phase sampling approach, however, allows to obtain posterior probabilities by applying (3.4.7), provided we are prepared to specify the posterior probabilities of the models after the first sample which are identical to the prior probabilities of the models before the second sample. Since these probabilities cannot be computed, they have to be chosen by means of subjective methods. A simplifying feature is that we need specify only one parameter per model in a subjective way.

The computations are carried out in such a way that we consider successive observations from the second sample and compute the posterior probabilities after each observation. This is done to gain insight into the development over time of the posterior probabilities, as well as the influence that a single observation can have on the results.

Most of the models considered were either already existent in the literature or are more or less obvious variations on known models. First, the demand for spirits in the first period was explained by real income per head and the deflated price of spirits. This result was studied extensively by means of partial scatter diagrams and regression diagrams describing the actual and explained consumption per head and the explanation obtained by the separate explanatory variables. This study mainly suggested that delayed reactions had to be incorporated in the models. Furthermore attention was paid to the order of magnitude and the algebraic signs of the parameters β, the Durbin-Watson statistic, and the sum of squared residuals λv. On the base of these investigations we selected twelve models. In the next section we shall present a description of the models studied.

6.2. DESCRIPTION OF THE MODELS

For the description of the models the following symbols are used throughout. The quantity of spirits per head is denoted by Q^*, real income per head by Y^*, and the deflated price of spirits by P^*. Their logarithms are denoted by q, y, and p, respectively. The first (backward) differences of these variables are denoted by $\Delta q(t)$, $\Delta y(t)$, and $\Delta p(t)$ $(t = 2, \ldots, 69)$, where $\Delta q(t) = q(t) - q(t-1)$ and the other differences are defined in an analogous way.

Each model contains a disturbance. We assume that this disturbance is normally distributed with zero mean and that, given a model, the disturbances are independent over time, such that their variance matrix is $h^{-1}I$.

Model I

The first model is a simple static constant-elasticity demand function. It is similar to that used in Section 4.4, the only difference being that the variables are first differences. It reads as follows

$$\Delta q(t) = \beta_1 \Delta y(t) + \beta_2 \Delta p(t) + \varepsilon_1(t) \tag{2.1}$$

Model II

$$\Delta q(t) = \beta_1 \Delta y(t) + \beta_2 \Delta y(t-1) + \beta_3 \Delta p(t) + \beta_4 \Delta p(t-1) + \varepsilon_2(t) \tag{2.2}$$

This[1] dynamic demand function expresses the idea of the delayed reaction of consumers on the changes of their income and the prices in such a way, that the adaptation to the new situation takes between one and two years.

Model III

$$\Delta q(t) = \beta_1 x_1(t) + \beta_2 x_2(t) + \beta_3 x_3(t) + \beta_4 x_4(t) + \varepsilon_3(t) \tag{2.3}$$

where

$$\begin{array}{lll} x_1(t) = \Delta y(t) & x_2(t) = 0 & \text{if } \Delta y(t) \geqslant 0 \\ x_1(t) = 0 & x_2(t) = \Delta y(t) & \text{if } \Delta y(t) < 0 \\ x_3(t) = \Delta p(t) & x_4(t) = 0 & \text{if } \Delta p(t) \geqslant 0 \\ x_3(t) = 0 & x_4(t) = \Delta p(t) & \text{if } \Delta p(t) < 0 \end{array}$$

We thus assume that there is a constant elasticity of demand to upward movements of income, and a constant elasticity with respect to downward movements, but that the two elasticities may differ; the assumption for the price elasticities is analogous. A further discussion of these so-called irreversible demand functions can be found in Farrell (1952).

Model IV

$$\Delta q(t) = \beta_1 \Delta y(t) + \beta_2 g_1(t) + \beta_3 \Delta p(t) + \beta_4 g_2(t) + \varepsilon_4(t) \tag{2.4}$$

where $g_1(t) = (t-1)y(t) - (t-2)y(t-1)$ and

1. We assume that β_3 and β_4 are not equal to zero to avoid that model I is a particular case of model II. In the remainder of this section similar assumptions will be made tacitly. Compare the footnote of Section 3.3.

$g_2(t)=(t-1)p(t)-(t-2)p(t-1)$. The systematic part of this model is obtained by applying the first-differences operator to

$$q(t)=\beta_1 y(t)+(t-1)\beta_2 y(t)+\beta_3 p(t)+(t-1)\beta_4 p(t)$$

where it is assumed that the elasticities change linearly over time. The income and price elasticities are $\beta_1+(t-1)\beta_2$ and $\beta_3+(t-1)\beta_4$, and it can easily be seen that the annual increments of these elasticities are β_2 and β_4, respectively.

Model V

$$\Delta q(t)=\beta_1 \Delta y(t)+\beta_2 \Delta y(t-1)+\beta_3 \Delta y(t-2)+\beta_4 \Delta p(t)+ +\beta_5 \Delta p(t-1)+\beta_6 \Delta p(t-2)+\varepsilon_5(t) \tag{2.5}$$

In this dynamic demand function it is assumed that the reaction to changes in income and prices takes still more time than in model II.

Model VI

$$\Delta q(t)=\beta_1 \Delta y(t)+\beta_2 \Delta p(t)+\beta_3 \Delta s_1(t)+\beta_4 \Delta s_2(t)+\varepsilon_6(t) \tag{2.6}$$

where $s_1(t)=\max\{y(t-1), y(t-2), y(t-3), y(t-4)\}$ and $s_2(t)=\min\{p(t-1), p(t-2), p(t-3), p(t-4)\}$. This model[1] expresses the idea that apart from current income and price the highest past income and lowest past price play a role. The past is defined as four years preceding the year of observation.

Model VII

$$\Delta q(t)=\beta_1 \Delta y(t)+\beta_2 \Delta p(t)+\beta_3 \Delta q(t-1)+\varepsilon_7(t) \tag{2.7}$$

Here the quantity consumed in the period $t-1$ is an explanatory variable for the quantity consumed in period t. This relation expresses in a way the assumption of habit formation. Note the similarity of this model with the partial-adaptation interpretation of the distributed-lag model as discussed in Appendix C. For simplicity, some more restrictive assumptions are made in (2.7).

Model VIII

This model combines the features of model III and model VII. The variables

1. This model is a variation of model IX discussed below.

x_1, x_2, x_3, and x_4, are defined as in model III.

$$(2.8) \qquad \Delta q(t) = \beta_1 x_1(t) + \beta_2 x_2(t) + \beta_3 x_3(t) + \beta_4 x_4(t) + \\ + \beta_5 \Delta q(t-1) + \varepsilon_8(t)$$

Model IX

$$(2.9) \qquad \Delta q(t) = \beta_1 \Delta y(t) + \beta_2 \Delta p(t) + \beta_3 \Delta u(t) + \varepsilon_9(t)$$

where $u(t) = \max \{q(t-1), q(t-2), q(t-3)\}$.

This model assumes more rigid habits in the consuming behavior. This model is discussed by Koyck (1956), and based on the work of Modigliani (1949).

The next three models are based on the so-called 'Expenditure Allocation Model' (or indirect addilog model) which was developed by Leser (1941-1942). An important theoretical property of this model is that it satisfies the adding-up criterion. A more detailed treatment of this model was given by Somermeyer and Wit (1956), and applications thereof can be found in Houthakker (1958) and Somermeyer, Hilhorst and Wit (1962). This allocation model is derived for individual households and M different commodities. It reads as follows

$$(2.10) \qquad U_s(t) = \frac{c_s(t)\{P_s(t)/Y(t)\}^{\alpha_s} Y(t)}{\sum_{r=1}^{M} c_r(t)\{P_r(t)/Y(t)\}^{\alpha_r}}$$

where U_s denotes expenditure to commodity s, Y income, P_s the price of commodity s, the subscripts s and r refer to the commodities 1, ..., M; c_s and c_r are called preference indicators (or preference coefficients), and t denotes the time period. We assume that this model can be applied to the 'typical consumer' and interpret U as expenditure per head and Y as income per head. We distinguish between two commodities, spirits (s) and non-spirits or remainder (r). Then, it follows from (2.10) that

$$(2.11) \qquad \frac{U_s(t)}{U_r(t)} = \frac{c_s(t)\{P_s(t)/Y(t)\}^{\alpha_s}}{c_r(t)\{P_r(t)/Y(t)\}^{\alpha_r}}$$

By definition we have $U_s(t) = P_s(t) Q_s(t)$, where Q_s denotes the quantity of spirits, and $Y(t) = U_s(t) + U_r(t)$. If we define

$$(2.12) \qquad q_s(t) = \log Q_s(t)$$

it follows that

$$(2.13) \qquad \Delta q_s(t) - \Delta q_r(t) = \\ \Delta \log c_s(t) - \Delta \log c_r(t) + (\alpha_s - 1) f_1(t) + (1 - \alpha_r) f_2(t)$$

where $\Delta q_s(t)$ is identical to $\Delta q(t)$ in the previous models and where
$f_1(t) = \log\{P_s(t)/Y(t)\} - \log\{P_s(t-1)/Y(t-1)\}$ and
$f_2(t) = \log\{P_r(t)/Y(t)\} - \log\{P_r(t-1)/Y(t-1)\}$.

From a statistical point of view (2.13), which is an ordinary double-logarithmic expression, is evidently more easily tractable than (2.10), which can not be linearized. To complete the model, a disturbance is added to (2.13). It has to be noted that some additional statistical data such as q_r and P_r are needed. They are presented (or can be computed from data) in Prest (1954), pp. 79, 86, and 175.

For the models X, XI, and XII some alternative assumptions are made about the preference indicators. More precisely, the last three models differ by alternative assumptions concerning these preference indicators.

Model X

In this model it is assumed that the preference indicators are constant. Then the model reads as follows

$$(2.14) \qquad \Delta q_s(t) - \Delta q_r(t) = \beta_1 f_1(t) + \beta_2 f_2(t) + \varepsilon_{10}(t)$$

where $\beta_1 = \alpha_s - 1$ and $\beta_2 = 1 - \alpha_r$.

Model XI

In this model it is assumed that the preference indicators are functions of Q, such that

$$(2.15) \qquad c_s(t) = \bar{c}_s\{Q_s(t-1)\}^{\delta_s}; \qquad c_r(t) = \bar{c}_r\{Q_r(t-1)\}^{\delta_r}$$

and that $\delta_s = \delta_r$ implying identical influences of lagged quantities of spirits and remainder. Combining (2.15) and $\delta_s = \delta_r$ with (2.13), we obtain

$$(2.16) \qquad \Delta q_s(t) - \Delta q_r(t) = \beta_1 f_1(t) + \beta_2 f_2(t) + \beta_3\{\Delta q_s(t-1) - \Delta q_r(t-1)\} + \varepsilon_{11}(t)$$

where $\beta_3 = \delta_s = \delta_r$.

Model XII

This model differs from model XI in the assumption that $\delta_s \neq \delta_r$. It can be written as

$$(2.17) \qquad \Delta q_s(t) - \Delta q_r(t) = \beta_1 f_1(t) + \beta_2 f_2(t) + \beta_3 \Delta q_s(t-1) + \beta_4 \Delta q_r(t-1) + \varepsilon_{12}(t)$$

6.3. ANALYSIS OF THE FIRST SAMPLE

This section deals with the statistical analysis of the models (discussed in Section 6.2) by means of the first sample. Since all models are defined in first-differences of logarithms, we assume the disturbances to be homoskedastic. Moreover, we assume that for all models we consider the disturbances to be normally distributed with zero mean and variance matrix $\sigma^2 I$.

The likelihood functions of the twelve models are combined with uniform prior densities (see Section 2.6) in order to obtain posterior densities. The numerical results given the first sample (the observations 1, ..., 34) can be found in Table 6.1. Note that no complete N matrices are given, but that we confined ourselves to presenting the determinants. The values of the Durbin-Watson statistics suggest a rather large amount of autocorrelation for most of the models. For the whole sample, however, the autocorrelation is reduced to about zero for all models under consideration. Therefore, we do not extend our set of models by taking into account a different number of autocorrelation parameters for the twelve models. The medians of β have been critically investigated. The signs and magnitudes appeared to be acceptable in the sense that they never contradicted our *a priori* thoughts or intuition, so that we decided not to incorporate other prior knowledge to obtain other (or better) results after the first sample. The values of λv and v show rather considerable differences in goodness of fit.

Now the problem is to assign the probabilities to the models given the first sample. Since a unique and objective way is not available, a subjective alternative has to be found. It is possible, for example, to order the models corresponding to their goodness of fit. We know, however, that in the uniform case other properties such as the determinants of the precision matrices and the limit of the ratio $\gamma'_{\alpha i}/\gamma'_{\alpha j}$ play an important role. Therefore, this method seems rather arbitrary and hardly acceptable. Other possibilities of solving this problem are to investigate the behavior of the estimated disturbances over time or the predictive power of the models. As already stated, all these methods are subjective and depend on the investigator concerned, so that no general rules can be advocated here. For our illustrative example we decided to start with equal prior probabilities[1] after the first sample and to investigate the consequences of this assumption for the final posterior probabilities. Note that, given our ultimate results, it is rather simple to compute the posterior probabilities for other prior probabilities; compare (3.4.7).

1. If one likes to compute the values of all lim $\gamma'_{\alpha i}/\gamma'_{\alpha j}$ for $\alpha \to \infty$, this is rather easy since they are implicitly defined, given p'_i $(i = 1, ..., 12)$. An unreasonable value of such a ratio could suggest how these prior probabilities could be changed.

TABLE 6.1. NUMERICAL RESULTS OF THE APPLICATION OF THE FIRST SAMPL

	I	*II*	*III*	*IV*	*V*	*VI*
n	34	33	34	34	32	33
r	2	4	4	4	6	4
$\|N\|$	0.34×10^{-4}	0.74×10^{-9}	0.24×10^{-10}	0.28×10^{-2}	0.13×10^{-13}	0.54×10^{-9}
λv	0.57×10^{-2}	0.39×10^{-2}	0.56×10^{-2}	0.53×10^{-2}	0.34×10^{-2}	0.42×10^{-2}
v	0.18×10^{-3}	0.14×10^{-3}	0.19×10^{-3}	0.18×10^{-3}	0.13×10^{-3}	0.15×10^{-3}
DW	0.93	1.18	0.97	1.01	1.08	1.07
$E\beta_1$	0.6321	0.2960	0.5884	0.7994	0.3770	0.5793
	(0.1829)	(0.1869)	(0.2396)	(0.3059)	(0.1950)	(0.1780)
$E\beta_2$	−0.6973	0.5455	0.9411	−0.0029	0.3518	−0.6051
	(0.1950)	(0.1819)	(0.4993)	(0.0016)	(0.2211)	(0.1821)
$E\beta_3$		−0.4390	−0.7077	−0.7091	0.2060	0.3442
		(0.1863)	(0.3021)	(0.2657)	(0.1892)	(0.2001)
$E\beta_4$		−0.5653	−0.6305	0.0009	−0.4512	−0.5561
		(0.1848)	(0.3182)	(0.0049)	(0.1998)	(0.1952)
$E\beta_5$					−0.3978	
					(0.2070)	
$E\beta_6$					−0.3013	
					(0.1924)	

6.4. COMPUTATION OF THE POSTERIOR PROBABILITIES

Up to now we collected nearly all the factors we need to compute the posterior probabilities. One thing needs some further discussion: the Jacobians of the transformations for the second part of the sample; compare Sections 3.3 and 3.4. It is clear that for the models I through IX the variable to be explained is the same. Then we can assign the value one to those Jacobians. For the models X, XI, and XII, however, some computations have to be done. Since these models have the same variable to be explained and the same assumptions concerning the disturbances and thus the same Jacobian with respect to the first nine models, we can confine ourselves to deriving the Jacobian of one of the models X, XI, XII with respect to one of the models I through IX. Then the variables to be explained are $\Delta q_s(t)-\Delta q_r(t)$ and $\Delta q_s(t)$, respectively, for $t=n_1+1, \ldots, n_1+n_2$; where n_1 (= 34) is the number of observations of the first sample and n_2 (= 34) the number of observations of the second sample. We assume that $q_s(n_1)$ and $q_r(n_1)$ are fixed and given since they have been observed in the first

…O THE TWELVE ALTERNATIVE MODELS[1]

VII	VIII	IX	X	XI	XII
33	33	33	34	33	33
3	5	3	2	3	4
0.27×10^{-6}	0.18×10^{-12}	0.21×10^{-6}	0.34×10^{-4}	0.29×10^{-6}	0.13×10^{-8}
0.27×10^{-2}	0.24×10^{-2}	0.46×10^{-2}	0.63×10^{-2}	0.47×10^{-2}	0.30×10^{-2}
0.09×10^{-3}	0.09×10^{-3}	0.16×10^{-3}	0.20×10^{-3}	0.16×10^{-3}	0.11×10^{-3}
2.35	2.66	1.02	0.93	1.71	2.33
0.5958	0.5639	0.7452	−0.6822	−0.5739	−0.4798
(0.1329)	(0.1644)	(0.1810)	(0.2041)	(0.1897)	(0.1548)
−0.5042	0.9793	−0.6998	1.0669	0.8026	0.9072
(0.1455)	(0.3404)	(0.1861)	(0.2181)	(0.2151)	(0.1754)
0.5901	−0.5732	0.4034		0.4004	0.6150
(0.1072)	(0.2072)	(0.1591)		(0.1363)	(0.1222)
	−0.3240				0.0096
	(0.2272)				(0.1501)
	0.6076				
	(0.1054)				

1. Note that the figures between parentheses correspond to the standard deviations of the relevant β. Some of them are rather large, but we disregard any of the consequences of such a property.

sample. Then the multivariate normal distributions of $q_s(t)-q_r(t)$ and $q_s(t)$ for $t=n_1+1, \ldots, n_1+n_2$ can be obtained. Note that the Jacobians of the transformations from the first differences to the levels are both equal to one, since the matrices of these transformations are triangular with all diagonal and lower left-hand elements equal to one. For simplicity, we compare the first-differences models via the corresponding level models. The Jacobian of the transformation (compare (3.3.2d)) now has the following form

$$(4.1) \qquad |J| = |\det [\partial \{q_s(t)-q_r(t)\}/\partial q_s(t')]|$$

for t, $t'=n_1+1,\ldots, n_1+n_2$. Because of

$$(4.2) \qquad Y(t)=P_s(t)Q_s(t)+P_r(t)Q_r(t)$$

we can find

$$(4.3) \qquad q_r(t)=\log [\{Y(t)-P_s(t)Q_s(t)\}/P_r(t)]$$

Combining (2.12) and (4.3), it is easy to verify that

$$(4.4) \qquad \partial\{q_s(t)-q_r(t)\}/\partial q_s(t') = 1+\frac{P_s(t)\,Q_s(t)}{P_r(t)\,Q_r(t)} \qquad \text{if } t=t'$$

$$= 0 \qquad \text{if } t<t'$$

and[1] that the value $|J|$ can be computed by

$$(4.5) \qquad |J| = \prod_{t=n_1+1}^{n_1+n_2}\left\{1+\frac{P_s(t)\,Q_s(t)}{P_r(t)\,Q_r(t)}\right\}$$

Now we have presented all elements required for computing the posterior probabilities. We computed them after each successive sample observation of the second sample. The results are presented in Table 6.2. For the observations during the first world war we assumed larger standard deviations on the disturbances. To be precise, for the period 1914-1918 we accepted a σ twice as large as for the other observations. We dropped the assumption of homoskedasticity for these few observations to take account of the possibility of large (random) shocks during this period.

Let us investigate the results presented in Table 6.2. The first row shows the prior probabilities all equal to $1/12 \approx 0.833333$. The models VII, VIII, and XII have in the beginning an impressive increase of their probability, but become rather quickly very improbable. Other models, viz., II, V, VI, and XI have, after a few observations, decreasing posterior probabilities and, furthermore, these probabilities finally diminish to a level much lower than the initial 1/12. A further investigation of the results shows remarkable jumps in the probabilities. The jumps in 1909 could be the result of the sudden price increase in that year. The price of spirits had been constant over 10 years and then increased immediately by 20 per cent. The answer to the question whether this jump actually took place, or that we are the victim of incorrect information about the prices at that time, is unknown to the author and outside the scope of the present study. In 1922 Ireland (apart from Ulster) left the United Kingdom after a war which started after the first world war. All the income and quantity data employed here include Ireland to 1919 and exclude it after that year. To reduce the whole sequences of data to a fairly comparable basis it was necessary to use population figures which include Ireland up to 1919 and exclude it afterward. This procedure is clearly not very satisfactory if the consumption per head in Ireland was remarkably different from that of the rest of the United

1. Note that for $t<t'$ the derivative (4.4) is zero since a change of $q_s(t')$ does not influence $q_s(t)$.

TABLE 6.2. POSTERIOR PROBABILITIES OF THE TWELVE MODELS AFTER SUCCESSIVE OBSERVATIONS: 1904-1938

Years	*I*	*II*	*III*	*IV*	*V*	*VI*
1904	.0833333	.0833333	.0833333	.0833333	.0833333	.0833333
1905	.0581436	.0672659	.0700956	.1041321	.0726947	.0449238
1906	.0310953	.0286402	.0433855	.0930929	.0337726	.0276599
1907	.0241872	.0153490	.0337907	.0692915	.0185271	.0142367
1908	.0237324	.0025701	.0375594	.0674111	.0026673	.0062629
1909	.0302920	.0716473	.0746045	.4144973	.1005184	.0047838
1910	.0424522	.0879108	.0825922	.4304007	.0811479	.0066573
1911	.0612062	.0124215	.1141235	.3887438	.0119893	.0094725
1912	.0494570	.0092271	.0968630	.4109026	.0083114	.0086059
1913	.0916569	.0191074	.1701098	.5012949	.0167120	.0166119
1914	.0923110	.0200158	.1604897	.5098156	.0170087	.0144214
1915	.1476999	.0229901	.1585178	.3428281	.0201099	.0213331
1916	.1540418	.0184680	.1533819	.3545318	.0184471	.0079547
1917	.1752012	.0211287	.1420338	.3148763	.0161900	.0085295
1918	.1644414	.0337240	.1379066	.3215003	.0248353	.0092222
1919	.1231038	.0037203	.1054090	.4294123	.0274535	.0048197
1920	.1841453	.0027944	.1351387	.3447959	.0034624	.0044538
1921	.2370954	.0032053	.1202515	.2381163	.0028175	.0021622
1922	.2482296	.0016838	.1223795	.2077262	.0020603	.0022044
1923	.2380724	.0015317	.1098510	.2432491	.0013374	.0021339
1924	.2349789	.0012660	.1061925	.2634977	.0008285	.0019888
1925	.2200316	.0011523	.0985249	.2986001	.0005980	.0018342
1926	.2074725	.0010022	.0903852	.3378722	.0005050	.0017007
1927	.1839483	.0011183	.0785426	.3907592	.0008741	.0019513
1928	.1652395	.0007236	.0699455	.4515010	.0004631	.0016075
1929	.1662610	.0007148	.0691575	.4492533	.0004309	.0015830
1930	.1416652	.0005716	.0582007	.5235372	.0003268	.0013833
1931	.1428883	.0005699	.0573801	.5207003	.0003193	.0013498
1932	.1047142	.0004928	.0423387	.6399296	.0002388	.0009901
1933	.0981503	.0004605	.0391729	.6618067	.0002165	.0009286
1934	.0787060	.0003220	.0315958	.7179461	.0001608	.0008122
1935	.0635069	.0002414	.0255964	.7623758	.0001332	.0007472
1936	.0680929	.0002553	.0268264	.7511008	.0001359	.0007554
1937	.0674480	.0002443	.0248770	.7542905	.0001320	.0007328
1938	.0494030	.0001668	.0180483	.8227423	.0000975	.0004905

Kingdom. This argument and the first world war seem a possible cause of the fairly wild behavior of the posterior probabilities about these years. For some models we investigated the estimated disturbances for this period and that of 1909, and we observed some rather large values which were

TABLE 6.2. (Continued)

Years	VII	VIII	IX	X	XI	XII
1904	.0833333	.0833333	.0833333	.0833333	.0833333	.0833333
1905	.1190362	.1325445	.0633908	.0614087	.0723578	.1340056
1906	.1745767	.2147036	.0460488	.0335105	.0826731	.1908402
1907	.2091109	.2628054	.0378868	.0259254	.0715676	.2173212
1908	.1412838	.4003375	.0505354	.0252180	.0941377	.1482838
1909	.0178555	.0351479	.1228326	.0457422	.0271690	.0549089
1910	.0007910	.0018590	.1844429	.0640836	.0134813	.0041805
1911	.0003598	.0006437	.2945804	.0922184	.0119797	.0022605
1912	.0002482	.0004258	.3301980	.0748510	.0093671	.0015423
1913	.0004302	.0007087	.0267931	.1382802	.0153566	.0029377
1914	.0004797	.0007922	.0266037	.1390899	.0157364	.0032352
1915	.0001602	.0002687	.0460827	.2244941	.0143203	.0011944
1916	.0000202	.0000134	.0509290	.2356409	.0062077	.0003629
1917	.0000220	.0000100	.0481757	.2667265	.0067264	.0003792
1918	.0000412	.0000186	.0466053	.2526641	.0083043	.0007364
1919	.0000030	.0000015	.1098513	.1922351	.0039521	.0000377
1920	.0000040	.0000017	.0308027	.2887417	.0056313	.0000273
1921	.0000046	.0000013	.0163074	.3730560	.0069500	.0000318
1922	.0000043	.0000013	.0169869	.3915736	.0071326	.0000168
1923	.0000042	.0000011	.0177705	.3790798	.0069526	.0000155
1924	.0000043	.0000011	.0194561	.3653392	.0064341	.0000122
1925	.0000040	.0000011	.0208651	.3523826	.0059943	.0000112
1926	.0000037	.0000009	.0227900	.3326481	.0056091	.0000100
1927	.0000038	.0000010	.0331799	.3041927	.0054172	.0000111
1928	.0000034	.0000008	.0310813	.2744395	.0049864	.0000079
1929	.0000034	.0000008	.0318335	.2758228	.0049305	.0000078
1930	.0000028	.0000007	.0343991	.2356606	.0042452	.0000063
1931	.0000028	.0000007	.0352802	.2372941	.0042077	.0000062
1932	.0000025	.0000006	.0311765	.1766796	.0034307	.0000054
1933	.0000024	.0000005	.0305752	.1654715	.0032093	.0000051
1934	.0000018	.0000004	.0335181	.1342375	.0026950	.0000036
1935	.0000014	.0000003	.0357755	.1093760	.0022424	.0000029
1936	.0000015	.0000003	.0335709	.1169049	.0023520	.0000031
1937	.0000015	.0000003	.0343779	.1155924	.0022999	.0000029
1938	.0000010	.0000002	.0217592	.0856405	.0016480	.0000021

about six times the estimated σ. All in all, there is enough reason to doubt the reliability of the data or to incorporate other explanatory variables in the models considered. We do not worry about this now, however, since we are only interested in giving a numerical illustration.

Looking at the final results (in 1938) we conclude that most of the models have very small probabilities, and that only five models have a posterior probability of more than one per cent.[1] Model IV is the most probable one (0.82), followed by the models X (0.09) and I (0.05), and finally IX (0.02) and III (0.02). We recall that these probabilities are computed given equal prior probabilities after the first sample. The investigator can ask himself which changes would occur if the prior probabilities were chosen in an alternative way. This could be the case, for example, if he appreciates the theoretical backgrounds of one of the models considered, e.g., model X, since it has been developed under such conditions that elementary requirements of certain model builders are satisfied. In such a case, he has to specify other prior probabilities, and the new final results can easily be computed; compare (3.4.7). We shall come back to this subject in Section 6.6.

It is not our intention to give a very detailed discussion about the possibilities of assigning prior probabilities. It is clear that many possibilities are available, and that it is difficult to choose. It is, in our opinion, rather important for which purpose the models are used. We confine ourselves to the case where the models are used for predictive purposes. Then, as will be shown in the next section, the posterior probabilities give rise to some interesting possibilities.

Finally, some limited sensitivity analysis has been carried out. We discuss for a few models the influence on the posterior probabilities if we change the value of $\lambda' v'$. It has been discussed in Section 6.3 that the values of b' of the twelve models are not contradicting our *a priori* thoughts. Further, it is difficult to criticize the values of $|N|$ as presented in Table 6.1, so that a change of such a determinant based on subjective feelings can hardly be defended. Therefore, we decided to confine ourselves to changing $\lambda' v'$ (which is the λv of Table 6.1). We discuss the models I, VII, and IX. Model VII shows an amazingly good fit in the first sample (see Table 6.1). It is possible that the small value of $\lambda' v'$ causes the small final posterior probability of this model; compare (3.4.7). Model IX has attractive predictive properties (see the next section), and is the most probable model of the set of dynamic models (but less probable than the static model I). Therefore, we think it is interesting to investigate its sensitivity for changing $\lambda' v'$. The values of $\lambda' v'$ of the models I, VII, and IX, are 0.0057, 0.0027, and 0.0046, respectively. We raise the values of $\lambda' v'$ of the models VII and IX to the value of $\lambda' v'$ of model I in six equal steps. Table 6.3 presents the ratios of

1. Note that this set of most probable models includes the simpler models I, IX, and X. This result corresponds to the conclusions of THORNBER and GEISEL discussed in Section 5.2.

the resulting final posterior probabilities. The first column in this table represents the original situation, the last the situation in which $\lambda' v' = 0.0057$ for all three models. We find that these ratios are rather sensitive to changes in $\lambda' v'$, and that although model I remains preferable to model VII, model IX becomes more probable than model I for increasing values of $\lambda' v'$.

TABLE 6.3. RATIOS OF POSTERIOR PROBABILITIES FOR INCREASING VALUES OF $\lambda' v'$

	1	*2*	*3*	*4*	*5*	*6*	*7*
p_1''/p_7''	49403	6799	1320	338	106	40	15
p_1''/p_9''	2.27	1.63	1.14	0.84	0.60	0.45	0.33

6.5. ON THE PREDICTIVE POWER OF THE MODELS

In comparing different models for the purpose of prediction, an investigator will usually confront them with the same set of data. If one of these models yields only a slightly better record according to some criterion, the investigator will not be certain whether that model is actually superior or whether the slight quality difference could reasonably be attributed to sampling variation. In such a case he could try to check whether the difference is significant before he adopts that model. To avoid the difficulties associated with such a procedure, we confine ourselves to comparing the conditional point predictions of the twelve models as well as some other point predictions obtained by methods discussed below.

The procedure used is the following. Given t sample observations, we predict $E\Delta q(t+1)$ for each model we consider, where the expectation is taken over the state space of the parameters of that model. This prediction is made given the explanatory variables in the period $t+1$. We follow Zellner and Chetty (1965), who derived that given a mean square prediction error as loss function the mean of the distribution of the dependent variable will be an optimal point prediction, optimal in the sense of minimizing the expected loss. Of course, if the model is used in a decision problem with an economic rather than a statistical loss function, the investigator should use that loss function; see, e.g., Fisher (1962).

For the models X, XI, and XII, the point prediction is $E\{\Delta q_s(t+1) - \Delta q_r(t+1)\}$. This value, however, can be recalculated by means of the equality $Y(t+1) = P_s(t+1)Q_s(t+1) + P_r(t+1)Q_r(t+1)$, such that the prediction $E\Delta q_s(t+1)$ comes available, which is directly comparable with the predictions of the models I through IX.

We now return to our two-phase sampling procedure. As our starting point we predict $\Delta q(1905)$ given the first sample. Then we revise our posterior distribution by making use of the first observation of the second sample, and continue this procedure for the whole second sample. Thus, after each new observation the parameters β are reestimated and used for the next prediction. It is clear that with this procedure we obtain 34 point predictions for each of the models we consider.

As a means of analyzing the accuracy of the predictions, we use the root mean square prediction errors (RMSPE) as suggested by, e.g., Theil (1955). We define P_{it} which is the prediction of the t-th period by means of the i-th model and A_t the corresponding actual outcome (or realization). Then the RMSPE is defined[1] as

$$(5.1) \qquad \text{RMSPE}_i = \sqrt{n_2^{-1} \sum_{t=n_1+1}^{n_1+n_2} (P_{it} - A_t)^2} \qquad \text{for } i = 1, \ldots, k$$

This coefficient is bounded below by zero. This level is attained only in the case of perfect predictions. On the other hand, rather large values can be obtained for the case of bad predictions.

Table 6.4 shows the computed values of the RMSPE for the twelve models, as well as the values obtained by three other methods:

(i) the predictions for a certain period $t+1$ are weighed with the posterior probabilities at time t. We call this method XIII;

(ii) the prediction for period $t+1$ is made by means of the model with the best fit (i.e., the smallest v'') at time t. We call this method XIV;

(iii) the prediction for period $t+1$ is made by means of the model with the largest posterior probability at time t. We call this method XV.

Note the fairly small differences of the RMSPE in Table 6.4. Most of the values lie between 0.026 and 0.035. Note also that the simple static models I and X, both of which have two explanatory variables only, predict rather well if compared with models containing more explanatory variables. If we consider the subset of dynamic models (that is, model II, V, VI, VII, VIII, IX, XI, and XII), the best result is obtained for model IX. If we order the posterior probabilities, as well as the RMSPE of the twelve models I-XII, to their magnitudes, we find that there is no monotonic relationship, although most of the deviations are not impressive; see Figure 3.

1. The predictions of the period 1914-1918 are treated in the same way as the other predictions.

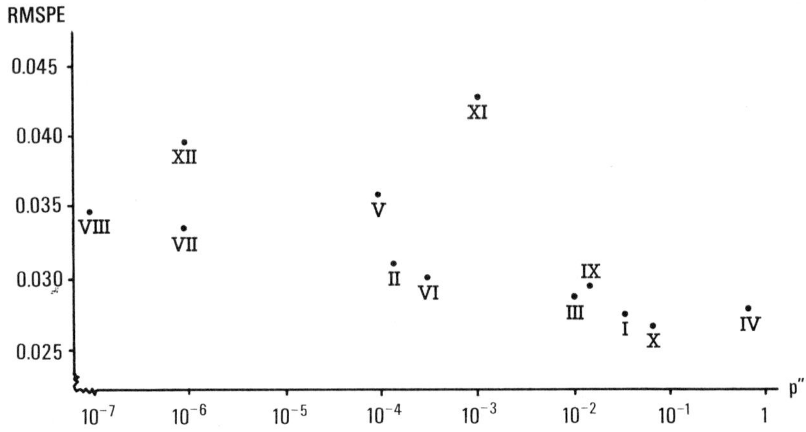

Fig. 3. *Posterior probabilities and* RMSPE *of the models I-XII*

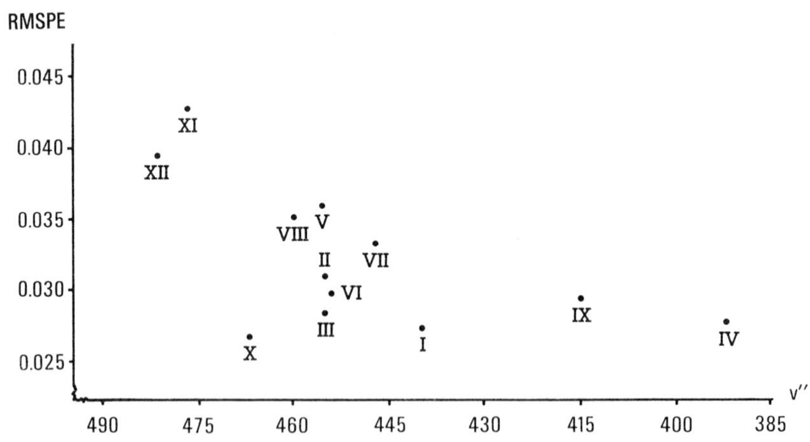

Fig. 4. v'' ($\times 10^6$) *and* RMSPE *of the models I-XII*

TABLE 6.4. RMSPE OF THE METHODS I–XV

Method	I	II	III	IV	V
RMSPE	0.0279	0.0314	0.0288	0.0275	0.0369
Method	VI	VII	VIII	IX	X
RMSPE	0.0302	0.0329	0.0349	0.0292	0.0271
Method	XI	XII	XIII	XIV	XV
RMSPE	0.0428	0.0396	0.0274	0.0326	0.0268

We note, however, that it is difficult to investigate the influence of rounding errors on the results of Table 6.4. If we order the final values of v'' and the RMSPE of these models,[1] see Figure 4, then we find a more untidy picture with a less smooth behavior than Figure 3.

From Table 6.2 we find that model IV is more probable than model X, while the RMSPE of model IV is larger than the RMSPE of model X; see Table 6.4. For this result, rounding errors for instance, as well as the choice of the prior probabilities can be responsible. From Table 6.4 we also find that method XIII, that is the predictions based on I through XII weighed with their most recent posterior probabilities, shows a RMSPE whose value lies between those of IV and X. A larger deviation is found if one applies method XIV, that is, using the model which has the smallest v'' at the moment of the prediction. This RMSPE is rather large in comparison with the RMSPE of the more probable models and method XIII. Method XV, on the other hand, generates the smallest RMSPE. We conclude therefore, that in this spirits example the procedure of posterior probabilities provides better results than the traditional $\bar{R}^2$ if the criterion of smallest RMSPE is accepted.[2]

6.6. SOME FURTHER NUMERICAL RESULTS OF THE TWO-PHASE SAMPLING APPROACH TO THE SPIRITS EXAMPLE

In this section some further investigations are carried out in order to gain a more detailed impression of the previous results, as well as to discuss the

1. If we order the $\bar{R}^2$ and the RMSPE of these models a similar picture is obtained.
2. If one likes, one can obtain smaller values of the RMSPE by changing the prior probabilities. If one expresses subjective ideas about the prior probabilities by means of linear inequality constraints, one can apply a mathematical programming method to obtain the smallest RMSPE. Of course, a sum constraint and twelve nonnegativity constraints have to be taken into account.

comparison of linear versus loglinear models. We start with this last subject. Suppose we like to compare two models, viz.,

$$(6.1) \qquad y = X\beta_1 + \varepsilon_1 \qquad \varepsilon_1 \sim N(0, h_1^{-1} I)$$

and

$$(6.2) \qquad z = W\beta_2 + \varepsilon_2 \qquad \varepsilon_2 \sim N(0, h_2^{-1} I)$$

with $z(t) = \log y(t)$ and $w_i(t) = \log x_i(t)$ for $t = 1, \ldots, n$, $i = 1, \ldots, r$, and w_i and x_i denote the i-th column of W and X, respectively.[1] Then the two-phase sampling approach immediately generates the prior and posterior values which are needed to compute the posterior probabilities. Furthermore the Jacobian of the transformation can be computed straightforwardly, since it is defined as

$$(6.3) \qquad |J| = |\det [\partial \log y(t)/\partial y(t')]|$$

for $t, t' = n_1+1, \ldots, n_1+n_2$. We applied this procedure to compare model I with its corresponding antilog model,[2] defined as

$$(6.4) \qquad \Delta Q^*(t) = \beta_1 \Delta Y^*(t) + \beta_2 \Delta P^*(t) + \varepsilon_{16}(t)$$

and call this model XVI. A previously mentioned advantage (see Section 6.4) of the two-phase sampling procedure is that the comparison of first-differences models is rather simple if the comparison of the corresponding level models is straightforward. The last observation of the first sample is then considered as given and fixed, so that the Jacobian of the transformation from level to first differences is equal to one. We computed the ratio of the posterior probabilities of the models I and XVI and obtained $p_1''/p_{16}'' = 7.3$. Since this value is considerably greater than one, we decided that further investigation of the antilog models is not very interesting.

Next, we investigated whether the static model IV, which appeared to be the most probable up to now, could be improved upon by adding a lagged variable. The first sample indicated already that $\Delta q(t-1)$ is a reasonable explanatory variable for $\Delta q(t)$. Therefore, we added this variable to model IV, and called this model XVII. The ratio of the posterior probabilities of the models IV and XVII is $p_4''/p_{17}'' = 506972$. So model IV

1. If (6.1) contains a constant term, all elements of the column of X corresponding to that constant term, are set equal to e or 10, and not equal to 1.
2. To be more precise, the level model corresponding to model XVI is obtained by applying the antilog transformation to the variables of the level model corresponding to model I.

is much more probable, while on the other hand, model XVII shows a better fit: the sums of the squared estimated disturbances $\lambda'' v''$ are 0.0292 and 0.0245 for model IV and model XVII, respectively. (This implies that the $\bar{R}^2$ of model IV is smaller than the $\bar{R}^2$ of model XVII).

Further, we compare the two most probable models, viz., IV and X. Model IV is defined as a model with linearly changing elasticities. Model X also generates elasticities which are variable over time,[1] although this variation is small. We computed the income and price elasticities given the whole sample, and for model X these values vary in the intervals (0.616, 0.608) and (−0.789, −0.784), respectively. The changes of these elasticities of model IV pass through the intervals (0.782, 0.870) and (−0.580, −0.924), respectively; so they are in absolute value considerably larger than the changes occurring in the elasticities generated by model X.

Finally, we investigate the sensitivity of the posterior probabilities of the models I through XII as to alternative assumptions concerning the heteroskedasticity of the disturbances in the second sample. We already discussed in Section 6.4 the sudden price increase in 1909, the first world war, the year 1919, and their possible influence on the posterior probabilities as presented in Table 6.2. Observing this table we also find other jumps, e.g. in 1905, 1913, 1927, and 1932 which we cannot explain historically. We investigated the behavior of λv, the elements of b, and the disturbances at the moment that such an observation was obtained, and we found rather impressive jumps in the values of λv and b, and very large disturbances. Therefore, we introduce another approach. First, we drop all previously made heteroskedastic assumptions. Then, given t sample observations, we predict $E\Delta q(t+1)$ for each of the twelve models, compute the prediction errors, divide these prediction errors by their corresponding estimated standard deviations, and weigh the so-computed elements with the posterior probabilities at time t. So we obtain the value

$$\text{(6.5)} \qquad \omega_t = \sum_{i=1}^{12} p''_{it} \{(P_{i,t+1} - A_{t+1})/v''_{it}\}$$

compare (5.1). Now we introduce the following decision rule; if $|\omega_t|$ is larger than some previously specified value ψ, the observation $t+1$ is divided by $|\omega_t/\psi|$, such that heteroskedasticity (a $|\omega_t \sigma/\psi|$ standard deviation) is accepted for that period. Note that such a procedure is rather arbitrary since observations which cannot be predicted by means of our models, can

1. Note that the elements of β of model X have not the interpretation of elasticities. The elasticities depend on these elements and are easy to compute. Compare SOMERMEYER, HILHORST, and WIT (1962).

be ruled out if $|\omega_t/\psi|$ is a large number. On the other hand, if one accepts that all models in the set considered are reasonable models, this procedure enables the investigator to assign a small weight to those observations which are probably influenced by exceptional events. Therefore, it is interesting to count the number of such corrections during the observation period.

Table 6.5 presents the final posterior probabilities of the 12 models for different values of ψ, viz., 3, 2, 1.8, and 1.5. In this table the symbols s and d denote static and dynamic models, respectively. The third column repeats the final results of Table 6.2. If we compare the new results with the results in this column, we find that the dynamic models (except V) become more probable than in the original case for at least one value of ψ; model XI even shows a large increase in probability. Model X, however, remains very probable, and model IV is still the most probable model for the cases we considered. Further, the number of corrections is larger the smaller the value of ψ. It is interesting to note that for all cases corrections have been made in the years 1905, 1909, 1910, 1915, 1916, 1918, 1919. Other already mentioned years in which corrections have been made are 1913, 1927, and 1932.

Finally, we investigate the sensitivity of this procedure for changing the prior probabilities after the first sample. Table 6.6 combines different prior probabilities with the heteroskedastic approach ($\psi = 2$) described in the previous paragraph. Under these circumstances the changes in the posterior probabilities resulting from the changed prior probabilities are less simple to compute than in the approach of Section 6.4. The first column of posterior probabilities in Table 6.6 has been computed for equal prior probabilities. In the second and third column it has been accepted that the dynamic models are 5 and 10 times, respectively, as probable *a priori* as the static models. In the columns four and five, finally, it has been accepted that the models X through XII, i.e., the models which satisfy the adding-up criterion, are 5 and 10 times, respectively, as probable *a priori* as the other models.

Investigating the results of Table 6.6 we find in the second column a remarkable increase in probability of the dynamic models. If one computes a new ratio of posterior probabilities of a dynamic and a static model, one finds that such a ratio is much greater than 5 times the corresponding ratio of the original result in the first column. Note, however, that model IV is still the most probable model. For the third column similar conclusions can be drawn, but now model XI is preferred to model IV. The columns four and five show, of course, increases in probability of the models X through XII. If one computes a new ratio of posterior probabilities of one of these models and one of the models I through IX, one does not find a value which differs much of the original ratio times the now accepted ratio of prior

TABLE 6.5. POSTERIOR PROBABILITIES OF THE TWELVE MODELS FOR DIFFERENT ASSUMPTIONS CONCERNING THE HETEROSKEDASTICITY

Model	*s/d*	2σ *1914–1918*	$\psi = 3$	$\psi = 2$	$\psi = 1.8$	$\psi = 1.5$
I	*s*	.0494030	.0476604	.0245244	.0226156	.0167251
II	*d*	.0001668	.0001762	.0000802	.0000675	.0000571
III	*s*	.0180483	.0172906	.0157983	.0165650	.0162907
IV	*s*	.8227423	.4562423	.6708079	.6623599	.5888624
V	*d*	.0000975	.0000318	.0000255	.0000257	.0000282
VI	*d*	.0004905	.0003298	.0004989	.0005417	.0005502
VII	*d*	.0000010	.0000012	.0000021	.0000023	.0000028
VIII	*d*	.0000002	.0000002	.0000008	.0000011	.0000023
IX	*d*	.0217592	.0429162	.0357707	.0383556	.0361403
X	*s*	.0856405	.4242034	.2015765	.1798598	.1243849
XI	*d*	.0016480	.0111375	.0508785	.0795532	.2168262
XII	*d*	.0000021	.0000099	.0000355	.0000521	.0001294
Number of corrections		—	7	8	10	15

TABLE 6.6. POSTERIOR PROBABILITIES OF THE TWELVE MODELS FOR $\psi = 2$ AND DIFFERENT PRIOR PROBABILITIES AFTER THE FIRST SAMPLE

Model	*s/d*	*All p′ equal*	*Dynamic versus Static* 5	10	*X–XII versus I–IX* 5	10
I	*s*	.0245244	.0095032	.0051773	.0112481	.0072372
II	*d*	.0000802	.0002259	.0003045	.0000353	.0000223
III	*s*	.0157983	.0100382	.0059651	.0088300	.0055791
IV	*s*	.6708079	.4615558	.2677799	.3859596	.2404638
V	*d*	.0000255	.0000744	.0001101	.0000097	.0000061
VI	*d*	.0004989	.0015831	.0019539	.0002581	.0001644
VII	*d*	.0000021	.0000095	.0000139	.0000010	.0000006
VIII	*d*	.0000008	.0000065	.0000106	.0000005	.0000003
IX	*d*	.0357707	.1574133	.2006884	.0194036	.0124229
X	*s*	.2015765	.0783219	.0425616	.4683345	.6025513
XI	*d*	.0508785	.2810604	.4750860	.1058393	.1314525
XII	*d*	.0000355	.0002072	.0003482	.0000798	.0000990
Number of corrections		8	8	8	8	8

probabilities. The corrections, 8 for each computation, have been made in the same years as for the case $\psi = 2$ in the previous example.

Some additional remarks can be made. First, model IV, i.e. the model with linearly changing elasticities, remained very probable for all cases we considered. Second, the dynamic models show a larger change in posterior probabilities for changing prior probabilities than the models which satisfy the adding-up criterion. Finally, we note that we confined ourselves to presenting the final results of the posterior probabilities. The behavior over time, however, did not show a markedly smoother pattern than in the original case (see Table 6.2) and is therefore not presented here.

6.7. CONCLUDING REMARKS

It is clear that experiments as have been carried out in this chapter can be continued *ad infinitum*. Model IV, for instance, showed a persistent large posterior probability in spite of our many attempts to find more probable models. This suggests, in our opinion, that an extension of the set of models considered with other models (including dynamic ones) with changing elasticities could be very interesting. And other than linear trends in the elasticities could be studied. All this, however, is beyond the scope of the present study.

As Goldberger (1968) has pointed out, investigators are wary of choosing a model which has a high $\bar{R}^2$ but parameters whose signs or magnitudes are not in accordance with their prior knowledge. It goes without saying that this behavior is rather subjective. The present study provides a general approach to the incorporation of such knowledge by means of a prior distribution, such that mutual comparison of alternative linear models is possible.

The application of posterior probabilities to models which differ only by the assumptions concerning the disturbances, is straightforward, and for the non-informative case even immediately applicable.

A general conclusion of this study, however, is that the practical worker has to be warned against the application of mechanical tools to the set of acceptable models. In this chapter it has been numerically illustrated that $\bar{R}^2$ and posterior probabilities do not always lead to the same conclusions.[1] Even in the illustrative case of 68 sample observations, which is a rather large sample in economic time-series analysis, the two-phase sampling approach

1. Unless the approaches of THORNBER (1966) and GEISEL (1970), as discussed in Chapter 5, have been accepted.

and $\bar{R}^2$ show large differences as to results. Usually, smaller samples are at the disposal of an investigator. Then it probably will occur again that the two-phase sampling method does not generate posterior probabilities such that the choice problem can be solved.[1] In such a case it seems advisable to postpone the choice problem and to go on with several models at the same time.

This seems to be the state of art: we have no reliable mechanism which makes unique choices, but we learn how to live with uncertainty as to the question which model is the best one. A positive feature of the present approach is that if the model builder's objectives are well defined, it yields a formal framework such that decisions or predictions can be made with a set of models which may be used either simultaneously or alternatively.

Another interesting result of this study, finally, is that it has demonstrated where and when in the analysis of comparing alternative linear models decisions have to be made in a subjective way.

1. Note that in case of a small sample, application of $\bar{R}^2$ is very dangerous; see Section 1.3.

Bibliography

ABRAHAMSE, A.P.J., and J. KOERTS (1969), 'New Estimators of Disturbances in Regression Analysis', Report 6906 of the Econometric Institute, Netherlands School of Economics. To be published in the *Journal of the American Statistical Association.*

ABRAHAMSE, A.P.J., and A.S. LOUTER (1969), 'On a New Test for Autocorrelation in Least-Squares Regression', Report 6920 of the Econometric Institute, Netherlands School of Economics. Published in *Biometrika*, 58, pp. 53–60 (1971).

ALMON, S. (1965), 'The Distribution Lag Between Capital Appropriations and Expenditures', *Econometrica*, 33, pp. 178–196.

BERTELS, K., and D. NAUTA (1969), *Inleiding tot het Modelbegrip*, De Haan, Bussum (The Netherlands).

BOX, G.E.P., and W.J. HILL (1967), 'Discrimination Among Mechanistic Models', *Technometrics*, 9, pp. 57–71.

CARNAP, R. (1962), *Logical Foundations of Probability*, The University of Chicago Press, Second Edition.

CHETTY, V.K. (1968), 'Bayesian Analysis of Haavelmo's Models', *Econometrica*, 36, pp. 582–602.

CHRIST, C.F. (1966), *Econometric Models and Methods*, John Wiley, New York.

COURANT, R. (1937), *Differential and Integral Calculus*, Vol. I, Blackie, London, Second Edition.

CRAMÉR, H. (1946), *Mathematical Methods of Statistics*, Princeton University Press, Princeton, N.J.

CRAMER, J.S. (1969), *Empirical Econometrics*, North-Holland Publishing Company, Amsterdam.

DHRYMES, P.J. (1969), 'Efficient Estimation of Distributed Lags with Autocorrelated Errors', *International Economic Review*, 10, pp. 47–67.

DURBIN, J., and G.S. WATSON (1950), 'Testing for Serial Correlation in Least-Squares Regression', *Biometrika*, 37, pp. 409–428.

DURBIN, J. (1960), 'Estimation of Parameters in Time-series Regression Models', *Journal of the Royal Statistical Society*, ser. B, 22, pp. 139–153.

FARRELL, M.J. (1952), 'Irreversible Demand Functions', *Econometrica*, 20, pp. 171–186.

FINETTI, B. DE (1937), 'La prévision: ses lois logiques, ses sources subjectives', *Annales de l'Institut Henri Poincaré*, 7, pp. 1–68.

FISHER, R.A., and F. YATES (1957), *Statistical Tables for Biological, Agricultural and Medical Research*, Oliver and Boyd, Edinburgh. Fifth revised edition.

Fisher, W.D. (1962), 'Estimation in the Linear Decision Model', *International Economic Review*, 3, pp. 1–29.

Geisel, M.S. (1970), 'Comparing and Choosing Among Parametric Statistical Models: A Bayesian Analysis with Macro-economic Applications', unpublished doctoral dissertation, University of Chicago.

Geisser, S. (1964), 'Posterior Odds for Multivariate Normal Classifications', *Journal of the Royal Statistical Society*, Series B, 26, pp. 69–76.

Goldberger, A.S. (1964), *Econometric Theory*, John Wiley & Sons, New York.

Goldberger, A.S. (1968), *Topics in Regression Analysis*, Macmillan, Collier-Macmillan, London.

Goldman, S. (1955), *Information Theory*, Prentice-Hall, New York. Third printing.

Griliches, Z. (1961), 'A Note on Serial Correlation Bias in Estimates of Distributed Lags', *Econometrica*, 29, pp. 65–73.

Groot, M.H. de (1970), *Optimal Statistical Decisions*, McGraw-Hill, New York.

Gupta, Y.P. (1969), 'Least Squares Variant of the Dhrymes' Two-Step Estimation Procedure of the Distributed Lag Model', *International Economic Review*, 10, pp. 112–113.

Harkema, R., and T. Kloek (1969), 'An Example of an Informative Prior Distribution in a Decision-Making Context'. Report 6913 of the Econometric Institute, Netherlands School of Economics.

Harkema, R. (1971), *Simultaneous Equations: A Bayesian Approach*, Rotterdam University Press, Rotterdam.

Heien, D.M. (1969), 'Income and Price Lags in Consumer-Demand Analysis', *Journal of the Royal Statistical Society*, Series A, pp. 265–271.

Hogg, R.V., and A.T. Craig (1965), *Introduction to Mathematical Statistics*, Second Edition, Macmillan, New York.

Houthakker, H.S. (1958), 'The Influence of Prices and Incomes on Household Expenditures', *Bulletin of the International Statistical Institute*, 37, Part 2, pp. 9–22.

Houthakker, H.S. (1960), 'Additive Preferences', *Econometrica*, 28, pp. 244–257.

Huang, D.S. (1970), *Regression and Econometric Methods*, John Wiley, New York.

Jeffreys, H. (1961), *Theory of Probability*, Oxford University Press, London, Third Revised Edition.

Johnston, J. (1963), *Econometric Methods*, McGraw-Hill, New York.

Jorgenson, D.W. (1966), 'Rational Distributed Lag Functions', *Econometrica*, 34, pp. 135–149.

Kendall, M.G., and A. Stuart (1961), *The Advanced Theory of Statistics*, Volume 2, Griffin, London.

Klein, L.R. (1958), 'The Estimation of Distributed Lags', *Econometrica*, 26, pp. 553–565.

Kloek, T. (1968), 'Een produktieproces in ontwikkeling', *De Economist*, 116, pp. 1–24.

Kloek, T. (1970), 'Note on Consistent Estimation of the Variance of the Disturbances in the Linear Model', Report 7003 of the Econometric Institute, Netherlands School of Economics.

Kloek, T., and F.B. Lempers (1970), 'Posterior Probabilities of Alternative Linear Models', Report 7006 of the Econometric Institute, Netherlands School of Economics.

Koerts, J. (1967), 'Some Further Notes on Disturbance Estimates in Regression Analysis', *Journal of the American Statistical Association*, 62, pp. 169–183.

Koerts, J., and A.P.J. Abrahamse (1969), *On the Theory and Applications of the General Linear Model*, Rotterdam University Press, Rotterdam.

KOYCK, L.M. (1954), *Distributed Lags and Investment Analysis*, North-Holland Publishing Company, Amsterdam.
KOYCK, L.M. (1956), *Verbruik en sparen*, Part 1, Edited by P.T.T., Tjeenk Willink & Zoon, Haarlem.

LEAMER, E.E. (1970), 'Distributed Lag Analysis with Informative Prior Distributions', Discussion Paper Number 147, Harvard Institute of Economic Research, Cambridge, Massachusetts.
LEEDE, E. DE, and J. KOERTS (1970), 'On the Notion of Probability: A Survey', Report 7007 of the Econometric Institute, Netherlands School of Economics.
LEMPERS, F.B., R. HARKEMA, and T. KLOEK (1969), 'On Vague Prior Information and Small Samples', Report 6911 of the Econometric Institute, Netherlands School of Economics.
LEMPERS, F.B., and T. KLOEK (1970), 'A Bayesian Procedure for Testing Regression Disturbances for Heteroskedasticity and Autocorrelation', Report 7016 of the Econometric Institute, Netherlands School of Economics.
LEMPERS, F.B., and A.S. LOUTER (1971), 'A Note on an Extension of the Table of the Student Distribution', Report 7101 of the Econometric Institute, Netherlands School of Economics. To be published in the *Journal of the American Statistical Association*, September 1971.
LESER, C.E.V. (1941-1942), 'Family Budget Data and Price Elasticities of Demand', *Review of Economic Studies*, pp. 40–57.
LESER, C.E.V. (1968), 'A Survey of Econometrics', *Journal of the Royal Statistical Society*, Series A, 131, pp. 530–566.
LINDLEY, D.V. (1965), *Introduction to Probability and Statistics from a Bayesian Viewpoint*, Cambridge University Press, London.
LIVIATAN, N. (1963), 'Consistent Estimation of Distributed Lags', *International Economic Review*, 4, pp. 44–52.

MALINVAUD, E. (1966), *Statistical Methods of Econometrics*, North-Holland Publishing Company, Amsterdam.
MALINVAUD, E. (1969), *Méthodes statistiques de l'économétrie*, seconde édition, Dunod, Paris.
MARSCHAK, J. (1953), 'Economic Measurements for Policy and Prediction', *Studies in Econometric Method*, ed. W.C. Hood and Tj.C. Koopmans, John Wiley, New York.
MODIGLIANI, F. (1949), 'Fluctuations in the Saving-Income Ratio: A Problem in Economic Forecasting', *Studies in Income and Wealth*, XI, pp. 371–443, National Bureau of Economic Research, New York.
MOOD, A.M., and F.A. GRAYBILL (1963), *Introduction to the Theory of Statistics*, McGraw-Hill Book Company, Inc., New York, Second Edition.
MUNROE, M.E. (1959), *Introduction to Measure and Integration*, Addison-Wesley Publishing Company, Massachusetts.

NERLOVE, M. (1958), 'Distributed Lags and Demand Analysis for Agricultural and Other Commodities', Washington: U.S. Department of Agriculture, Agricultural Handbook no. 141.

ORCUTT, G.H. (1967), 'Microeconomic Analysis for Prediction and National Accounts', *Forecasting on a Scientific Basis*, Centro de Economica E Finanças, Lisboa.

PEARSON, E.S., and H.O. HARTLEY (1956), *Biometrika Tables for Statisticians*, Volume I, Cambridge University Press. Fourth revised edition.
PLACKETT, R.L. (1966), 'Current Trends in Statistical Inference', *Journal of the Royal Statistical Society*, Series A, 129, pp. 249–267.
PRATT, J.W., H. RAIFFA, and R. SCHLAIFER (1964), 'The Foundations of Decision under Uncertainty: An Elementary Exposition', *Journal of the American Statistical Association*, 59, pp. 353–375.
PREST, A.R. (1949), 'Some Experiments in Demand Analysis', *The Review of Economics and Statistics*, 31, pp. 33–49.
PREST, A.R. (1954), *Consumers' Expenditure in the United Kingdom 1900-1919*, Cambridge University Press.

RAIFFA, H., and R. SCHLAIFER (1961), *Applied Statistical Decision Theory*, Graduate School of Business Administration, Harvard University, Boston.
RAIFFA, H. (1968), *Decision Analysis*, Addison-Wesley, Massachusetts.

SAVAGE, L.J. (1954), *The Foundations of Statistics*, Wiley, New York.
SAVAGE, L.J. (1962), Bayesian Statistics. *Recent Developments in Information and Decision Processes*, pp. 161–194, ed. by R.E. Machol and P. Gray, Macmillan, New York.
SCHEFFÉ, H. (1947), 'A Useful Convergence Theorem for Probability Distributions', *Annals of Mathematical Statistics*, Vol. 18, pp. 434–438.
SOLOW, R.M. (1960), 'On a Family of Lag Distributions', *Econometrica*, 28, pp. 393–406.
SOMERMEYER, W.H., and J.W.W.A. WIT (1956), 'Een verdeelmodel', Report M14 of the Netherlands Central Bureau of Statistics.
SOMERMEYER, W.H., J.G.M. HILHORST, and J.W.W.A. WIT (1962), 'A Method for Estimating Price and Income Elasticities from Time Series and its Application to Consumers' Expenditures in the Netherlands, 1949-1959', *Statistical Studies*, No. 13, pp. 30–53. Prepared by the Netherlands Central Bureau of Statistics.
SOMERMEYER, W.H. (1967), 'Specificatie van economische relaties', *De Economist*, 115, pp. 1–26.
STONE, R. (1954), *The Measurement of Consumer's Expenditure and Behaviour in the United Kingdom 1920-1938*, Cambridge University Press, London.

THEIL, H. (1955), 'Who Forecasts Best?' *International Economic Papers*, No. 5, pp. 194–199.
THEIL, H. (1957), 'Specification Errors and the Estimation of Economic Relationships', *The Review of the International Statistical Institute*, 25, pp. 41–51.
THEIL, H. (1958), *Economic Forecasts and Policy*, North-Holland Publishing Company, Amsterdam.
THEIL, H., and A.L. NAGAR (1961), 'Testing the Independence of Regression Disturbances', *Journal of the American Statistical Association*, 56, pp. 793–806.
THEIL, H. (1963), 'On the Use of Incomplete Prior Information in Regression Analysis'. *Journal of the American Statistical Association*, 58, pp. 401–414.
THEIL, H. (1965), 'The Analysis of Disturbances in Regression Analysis', *Journal of the American Statistical Association*, 60, pp. 1067–1079.
THEIL, H. (1966), *Applied Economic Forecasting*, North-Holland Publishing Company, Amsterdam.
THORNBER, H. (1966), 'Applications of Decision Theory to Econometrics', Unpublished Doctoral Dissertation, University of Chicago.

THORNBER, H. (1967), 'Finite Sample Monte Carlo Studies: An Autoregressive Illustration', *Journal of the American Statistical Association*, 62, pp. 801–818.
TIAO, G.C., and A. ZELLNER (1964), 'Bayes' Theorem and the Use of Prior Knowledge in Regression Analysis', *Biometrika*, 51, pp. 219–230.
TOULMIN, S. (1960), *The Philosophy of Science: An Introduction*, Harper and Row, New York.

ZELLNER, A., and G.C. TIAO (1964), 'Bayesian Analysis of the Regression Model with Autocorrelated Errors', *Journal of the American Statistical Association*, 59, pp. 763–778.
ZELLNER, A., and V.K. CHETTY (1965), 'Prediction and Decision Problems in Regression Models from the Bayesian Point of View', *Journal of the American Statistical Association*, 60, pp. 608–616.
ZELLNER, A., and C.J. PARK (1965), 'Bayesian Analysis of a Class of Distributed Lag Models', The Econometric Annual of the *Indian Economic Journal*, 13, pp. 432–444.
ZELLNER, A., and M.S. GEISEL (1968), 'Analysis of Distributed Lag Models with Applications to Consumption Function Estimation', Paper presented at the European Meeting on Statistics, Econometrics, and Management Science, Amsterdam 1968, University of Chicago. Published in *Econometrica*, 38, pp. 865–888 (1970).
ZELLNER, A. (1971), *Bayesian Inference in Econometrics*, to be published by John Wiley and Sons, Inc.

Index

(B 1141) Printed in Belgium by Ceuterick s.a.
Brusselse straat 153 3000-Louvain
Managing director L. Pitsi Bertemse baan 25 3008-Veltem-Beisem